T0134911

Springer Theses

Recognizing Outstanding Ph.D. Research

Aims and Scope

The series "Springer Theses" brings together a selection of the very best Ph.D. theses from around the world and across the physical sciences. Nominated and endorsed by two recognized specialists, each published volume has been selected for its scientific excellence and the high impact of its contents for the pertinent field of research. For greater accessibility to non-specialists, the published versions include an extended introduction, as well as a foreword by the student's supervisor explaining the special relevance of the work for the field. As a whole, the series will provide a valuable resource both for newcomers to the research fields described, and for other scientists seeking detailed background information on special questions. Finally, it provides an accredited documentation of the valuable contributions made by today's younger generation of scientists.

Theses are accepted into the series by invited nomination only and must fulfill all of the following criteria

- They must be written in good English.
- The topic should fall within the confines of Chemistry, Physics, Earth Sciences, Engineering and related interdisciplinary fields such as Materials, Nanoscience, Chemical Engineering, Complex Systems and Biophysics.
- The work reported in the thesis must represent a significant scientific advance.
- If the thesis includes previously published material, permission to reproduce this must be gained from the respective copyright holder.
- They must have been examined and passed during the 12 months prior to nomination.
- Each thesis should include a foreword by the supervisor outlining the significance of its content.
- The theses should have a clearly defined structure including an introduction accessible to scientists not expert in that particular field.

More information about this series at http://www.springer.com/series/8790

Svenja Karen Pflitsch

Associated Production of W + Charm in 13 TeV Proton-Proton Collisions Measured with CMS and Determination of the Strange Quark Content of the Proton

Doctoral Thesis accepted by
Deutsches Elektronen Synchrotron, Hamburg,
Germany

 Springer

Author
Dr. Svenja Karen Pflitsch
Particle Physics
Deutsche Elektronen-Synchrotron DESY
Hamburg, Germany

Supervisor
Dr. Katerina Lipka
Particle Physics
Deutsches Elektronen Synchrotron
Hamburg, Germany

ISSN 2190-5053 ISSN 2190-5061 (electronic)
Springer Theses
ISBN 978-3-030-52764-8 ISBN 978-3-030-52762-4 (eBook)
https://doi.org/10.1007/978-3-030-52762-4

This Springer imprint is published by the registered company Springer Nature Switzerland AG
The registered company address is: Gewerbestrasse 11, 6330 Cham, Switzerland

Supervisor's Foreword

The structure of the nucleon is one of the most fundamental topics of particle physics. The mass of the nucleon, and therefore of all nuclear matter in the universe, has its origin in the strong interaction of the elementary constituents of the nucleon—the quarks and the gluons. This interaction is described by the quantum field theory Quantum Chromodynamics (QCD). Studying the details of the proton structure gains highest attention in the context of the interpretation of the Standard Model measurements and searches for New Physics at the CERN LHC.

In QCD, the quantum properties of the nucleon are ascribed to the three valence quarks, while the energy of the gluons and the sea quarks is responsible for its mass. The proton structure is expressed in terms of universal Parton Distribution Functions (PDFs), representing probabilities to find a parton (quark or gluon) in the proton, carrying a fraction x of is momentum at a particular energy scale. While the scale dependence of the PDFs can be calculated in perturbation theory using QCD evolution equations, their x-dependence has to be extracted from the experimental measurements. The investigation of the flavour decomposition of the quark sea represents a major experimental challenge. In particular, the accurate knowledge of the strange quark content of the proton is crucial for determination of the electroweak parameters of the Standard Model in proton-proton collisions at the LHC. Before the LHC era, constraints on the strange-quark distribution were obtained from neutrino scattering experiments. These measurements probe the (anti)strange-quark density at $x \approx 10^{-1}$ and scales of approximately 10 GeV^2, but their interpretation is complicated by nuclear corrections and uncertainties in the charm-quark fragmentation function.

Production of electroweak bosons, W and Z, in proton-proton collisions at the LHC has complimentary sensitivity to the strange quark distribution at the scale of the boson mass, extending probed x to $\approx 10^{-3}$. In the last decade, indirect probes of strange quark at the LHC using inclusive W and Z boson production lead to controversial interpretations. While the experimental measurements are very precise, their QCD interpretation is nontrivial, since quark combinations of different flavours are probed. In contrast, the measurements of associated W + charm production in proton-proton collisions have great potential to access the strange-quark distribution directly through the leading order QCD process $g + s \rightarrow W + c$.

Dr. Pflitsch has performed the measurement of associated production of W + charm in proton-proton collisions at the LHC using the data collected by the CMS detector at a center-of-mass energy of 13 TeV. Quarks can't be observed as free particles, and the particular challenge of this work is the charm quark identification. In the work by Svenja Pflitsch, the charm quarks are tagged via their hadronisation into D* mesons, which are fully detected in decay channel $D^* \rightarrow D^0 + \pi$, based on the track reconstruction down to the transverse momenta of 0.5 GeV. This channel provides the most clean experimental signature. For the first time, it was possible to measure W + charm cross section precisely using the full meson reconstruction. Using the results of her measurements, Svenja Pflitsch has performed the global QCD analysis with the PDF determination having investigated different variants of the parameterization. This way, the strange content of the proton is probed directly, providing unambiguous results and resolving the controversy in the field.

The analysis strategy used by Svenja Pflitsch provides new insights into the flavour decomposition of the proton quark sea and is a major step towards better understanding of the matter structure.

Hamburg, Germany   Dr. Katerina Lipka
April 2020

Abstract

This thesis presents the measurement of associated production of a $W^\pm$ boson and a charm quark ($W + c$) in proton-proton collisions at a center-of-mass energy of 13 TeV. The data used in this analysis has been recorded by the CMS experiment at the CERN LHC and corresponds to an integrated luminosity of 35.7 fb^{-1}. The $W^\pm$ bosons are reconstructed by the presence of a muon and a neutrino, with the latter indicated by the missing transverse momentum in an event. Charm quarks are identified by the full reconstruction of $D^*(2010)^\pm$ mesons decaying via $D^*(2010)^\pm \rightarrow D^0 + \pi^\pm \rightarrow K^\mp + \pi^\pm + \pi^\pm$. The fiducial phase space of the measurement is defined by the muon transverse momentum $p_T^\mu > 26$ GeV, muon pseudorapidity $|\eta^\mu| < 2.4$ and the charm quark transverse momentum $p_T^c > 5$ GeV. The measurement is performed inclusively and differentially as a function of the absolute pseudorapidity of the muon from the $W^\pm$ boson decay. The results are compared to theoretical predictions using different PDF sets. A subsequent QCD analysis is performed to extract the strange quark content of the proton and assess possible improvements in the uncertainties associated with the distribution by including the new measurement. The extracted strange quark distribution is compared to distributions obtained in global PDF fits, which utilize the results of neutrino scattering experiments.

Acknowledgements

I want to thank my supervisor, Dr. Katerina Lipka, for giving me the opportunity to work on this interesting topic which had "something to do with Ws" and beyond. Thank you for the advice, discussions and support, which made it possible that the results of this thesis could already be presented at several conferences and were published in a scientific journal. I am grateful for the time you took to help me improve my research, as well as my presentation skills.

Thank you to Prof. Dr. Sven-Olaf Moch for reviewing this thesis and being part of my defence committee. I would also like to express my gratitude to the other members of my defence committee, Prof. Dr. Günter Sigl, Prof. Dr. Elisabetta Gallo and Dr. Hannes Jung.

Thank you to my office mates Engin Eren, Mykola Savitskyi and Till Arndt, who were always up for a good discussion, be it about physics, politics, or movies. Thank you to Benoit Roland for the many discussions about $W + c$ and Paolo Gunnellini for everything related to MC. I also want to thank my colleagues at DESY who always took the time to give advice when it was needed and generally creating a great working atmosphere. In this context, I would like to give a special thanks to the QCD group.

And finally, another special thanks to my family for their continuous moral support and patience.

Contents

Chapter 1
Introduction

Improving the understanding of the fundamental particles of matter and their interactions is one of the driving forces of physics research. For this purpose a variety of experimental measurements are conducted to test predictions based on the Standard Model of particle physics or search for novel phenomena. Despite advancements in the accuracy of theoretical calculations, important aspects of matter description, such as the proton (nucleon) structure, cannot be derived from first principles but have to be determined from experimental data. Theoretical predictions describing the interaction between quarks and gluons, are to a large extend calculated pertubatively (pQCD) via an expansion of the strong coupling (α_S). This provides an approximate solution to QCD at large energy transfers between the colliding partons, where α_S is sufficiently small.

The cross section of many processes observed in hadronic collisions can be calculated via QCD factorization, which is a convolution of the partonic cross section, determined in pQCD, and parton distribution functions (PDFs). PDFs are defined as the probability of a parton to carry the fraction x of proton momentum and are determined phenomenologically in QCD analyses. There, theoretical calculations at a fixed order of pQCD are quantitatively compared to the results of measurements, sensitive to different quark flavours inside the nucleon. QCD analyses that are using a large variety of datasets from many different experiments are generally referred to as *Global PDFs*. These are provided by dedicated PDF groups, such as ABMP, CTEQ, MMHT or NNPDF, with each group differing in the used datasets, statistical approaches or model assumptions. Due to the limited number of measurements directly sensitive to the strange quark content of the nucleon, the s-quark distribution is least constrained among the light quarks. Many results used to determine this distribution are from charged current deep inelastic scattering (DIS) in (anti)neutrino-iron experiments and probe strangeness via charm-dimuon production.

© The Editor(s) (if applicable) and The Author(s), under exclusive license
to Springer Nature Switzerland AG 2020
S. K. Pflitsch, *Associated Production of W + Charm in 13 TeV
Proton-Proton Collisions Measured with CMS and Determination of the Strange Quark
Content of the Proton*, Springer Theses,
https://doi.org/10.1007/978-3-030-52762-4_1

At hadron colliders, such as the Tevatron at Fermilab or the Large Hadron Collider (LHC) at CERN (*Conseil Européen pour la Recherche Nucléaire*), the measurement of the associated production of a W boson and a charm quark (W + c) may provide important information about the strange quark in the nucleon. The process is directly sensitive to the strange quark content of the proton, as it is predominantly produced by the hard scattering of a strange quark and a gluon ($g + \bar{s} \rightarrow W^+ + \bar{c} + X$) at the leading order of pQCD ($\mathcal{O}(\alpha_S)$). The data used for the measurements presented in this thesis were recorded at the LHC, at a centre-of-mass energy of $\sqrt{s} = 13$ TeV. This is the highest energy reached with a circular hadron collider to date, making it ideal for precision test of the Standard Model, as well as searches for new phenomena. The collision data is recorded with dedicated detectors, capable of measuring the trajectories and energy of the particles produced with high precision.

One of the detectors that are installed around the interaction points of the LHC, is the Compact Muon Solenoid (CMS). It is a general purpose detector, which can determine the energy and momentum of particles over a broad range with good accuracy. By combining the signals of different detector modules, particles such as muons or electrons can be identified with high purity. The first measurement of W + c at the LHC was performed by the CMS collaboration at a center-of-mass energy of $\sqrt{s} = 7$ TeV with a focus on c-jets, using different charm identification techniques using a data sample corresponding to $5\,\text{fb}^{-1}$.

In this thesis, the cross section of W + c is determined using the pp-collision data collected by the CMS experiment in 2016, at a center-of-mass energy of $\sqrt{s} = 13$ TeV, corresponding to an integrated luminosity of $35.7\,\text{fb}^{-1}$. The $W^\pm$ bosons are identified by their leptonic decay to a muon and a neutrino and the charm quarks are tagged via the full reconstruction of $D^*(2010)^\pm$ mesons via the decay $D^*(2010)^\pm \rightarrow D^0 + \pi^\pm \rightarrow K^\mp + \pi^\pm + \pi^\pm$. The measurement is performed in the fiducial phase space defined by the transverse momentum $p_T^\mu > 26$ GeV and pseudorapidity $|\eta^\mu| < 2.4$ of the muon and by the transverse momentum of the charm quark $p_T^c > 5$ GeV. The cross section is determined both inclusively and as a function of the absolute pseudorapidity of the muon from the $W^\pm$ boson decay and compared to theoretical calculations combined with different PDFs.

The results from the 13 TeV W + c analysis are subsequently used, in combination with other relevant CMS results and the most precise HERA DIS data, to perform a QCD analysis at next-to-leading order (NLO) to probe the strange quark content of the proton. The impact of the experimental uncertainties connected with the measurements on the extracted PDFs is evaluated using two different methods. The resulting strange quark distribution and suppression, with respect to the other light sea quarks, is compared to the distributions published by well-established PDF groups.

This thesis is organized as follows: An introduction to the Standard Model of particle physics with a focus on QCD factorization and the proton structure, is given in Chap. 2. The experimental setup of the LHC and the CMS detector is described in Chap. 3. The reconstruction of events from the detector data is explained in Chap. 4. The measurement of the W + c cross section in 13 TeV pp-collisions is presented in Chap. 5, and the interpretation of the results in terms of a QCD analysis to determine the strange quark content of the proton is shown in Chap. 6. A summary and discussion of the results is presented in Chap. 7.

Chapter 2
Theoretical Overview

This chapter presents the theoretical basics necessary for the extraction of the strange quark content of the proton, which is the main motivation for the measurement of the W + c production cross section. Here, the Standard Model of particle physics is presented with the forces it describes, the proton structure and measurements capable of probing it, as well as an overview on techniques used in theoretical predictions. Furthermore, the QCD analysis tools used to extract the proton structure, are discussed.

2.1 The Standard Model of Particle Physics

The Standard Model of particle physics describes the fundamental constituents of matter and their interactions. The elementary particles, called *fermions* (spin-$\frac{1}{2}$ particles) and *gauge bosons* (spin-1 particles) [1, 2], show different symmetry properties, according to their quantum numbers. While fermions make up all visible matter in the universe, bosons are the mediators of the strong, electromagnetic and weak interactions, as well as the interaction responsible for particle mass generation. Based on current knowledge, both particle types are considered elementary, meaning they do not have a substructure of their own [3]. The properties of the fundamental particles, such as charge, spin and mass are listed in Fig. 2.1.

Different types of fermions are referred to as *flavours* and are categorized into three generations consisting of two *quarks* and two *leptons* each, with the particle masses increasing with generations. A corresponding anti-particle with the same mass, but opposite charge exists for each type of fermion, and is conventionally denoted by an overbar. One of the quarks is of *up* type (up, charm, top), with a +2/3[1] charge,

---

[1] With respect to the elemental charge of an electron.

© The Editor(s) (if applicable) and The Author(s), under exclusive license
to Springer Nature Switzerland AG 2020
S. K. Pflitsch, *Associated Production of W + Charm in 13 TeV
Proton-Proton Collisions Measured with CMS and Determination of the Strange Quark
Content of the Proton*, Springer Theses,
https://doi.org/10.1007/978-3-030-52762-4_2

Standard Model of Elementary Particles

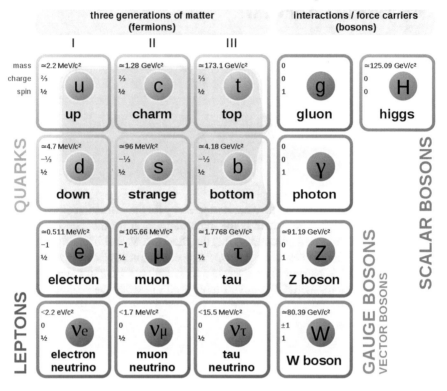

Fig. 2.1 Fundamental particles of the standard model with their quantum properties and masses [4]

and one of *down* type (down, strange, bottom), with a charge of $-1/3$. Lepton generations consist of one charged particle (e, μ, τ) and one corresponding neutrino (ν_e, ν_μ, ν_τ). The mass of the neutrino is the smallest among all fundamental particles and therefore difficult to determine, although it can be probed via neutrino oscillation measurements.

The mathematical formulation of the Standard Model follows the Lagrangian formalism, where all dynamics of a system are described by a single function. It is based on quantum field theory with a $SU(3) \times SU(2) \times U(1)$ gauge symmetry. The particles of the Standard Model are treated as quantum fields, represented by wave functions with different transformation symmetries, depending on their type. Fermions are antisymmetric under the exchange of two particles and their evolution in space-time is described by the Dirac equation, describing particles with spin $1/2$. The solution of this equation is invariant under the Lorentz transformations of special relativity and provides two solutions, one with positive and one with negative energy, with the negative solution being interpreted as the corresponding anti-particle.

Bosons are symmetric under the same transformation and their space-time evolution follows the Klein–Gordon equation for spin 1 particles.

Interactions between particles are often visualized by so called *Feynman diagrams*, in which the vertices represent the couplings of the interactions and antiparticles are shown as arrows pointing backwards in time. Open lines correspond to initial state or final state particles, whereas closed lines represent virtual particles that are created and reabsorbed. The boson mediating the reaction can be a virtual particle, meaning that it does not carry the same mass as the real particle and therefore cannot be observed.

The Standard Model describes the electromagnetic, weak and strong interactions, whose phenomenological and mathematical properties are illustrated in more detail in the following sections. To this date, the Standard Model does not include gravity. However, due to its relatively weak interaction strength between two elementary particles, the gravitational interaction is presumed to have no significant effect on the experimental results at current particle collision energies and can therefore be neglected.

2.1.1 The Electromagnetic Interaction

The electromagnetic interaction couples to all electrically charged particles, and is described by the theory of Quantum Electrodynamics (QED). It's mediator particle is the *photon*, a massless and neutral gauge boson. Since photons do not carry charge themselves, a self-coupling is not possible. This property, and the fact that photons are massless particles, leads to an infinite reach for the EM interaction, with its strength decreasing with the distance between the interacting particles.

The coupling strength of the EM interaction is given by the so called *fine structure constant*:

$$\alpha_{\mathrm{EM}} = \frac{e^2}{4\pi\epsilon_0 \hbar c} \approx \frac{1}{137} \, , \tag{2.1}$$

where e is the electric charge of the electron, and ϵ the vacuum permittivity. It's value increases at short distances or for interactions taking place at high energy scales. This effect is caused by vacuum fluctuations producing and re-absorbing virtual e^+e^- pairs that surround the interacting particles. Therefore, the *net-charge* experienced in the interaction deviates from the charge carried by the real particles alone. This property is also known as the *charge-screening* effect.

Mathematically, QED is an abelian gauge theory with a $U(1)$ symmetry and its Lagrangian density can be written as follows:

$$\mathcal{L}_{U(1)_{\mathrm{EM}}} = -\frac{1}{4\mu_0} F_{\mu\nu}^i F_i^{\mu\nu} + \bar{\psi}(i\gamma^\mu D_\mu - mc^2)\psi$$

with: $F_{\mu\nu} = \partial_\mu A_\nu - \partial_\nu A_\mu$

$$D_\mu = \partial_\mu - ieA_\mu \tag{2.2}$$

Here, ψ and $\bar{\psi}$ represent the quantum fields of the interacting charged particles. D_μ is the covariant derivative, γ^μ are the gamma-matrices and $F_{\mu\nu}$ is the electromagnetic tensor. The four-vector potential A^μ is chosen such that the *Lorentz condition* $\partial_\mu A^\mu = 0$ can be applied as a gauge fixing condition.

2.1.2 The Weak Interaction

All quarks and leptons are weakly interacting and in the case of neutrinos it is the only interaction they participate in. The weak interaction is mediated by three carrier particles of which two are charged (W$^+$, W$^-$) and one is neutral (Z^0). The particles have large masses with 80.4 GeV [5] for the W$^\pm$ bosons and 91.2 GeV for the Z^0 boson, leading to an extremely short interaction range of $\approx 2 \cdot 10^{-18}$ m.

The Z^0 boson mediates processes such as electron-neutrino scattering ($e^- + \nu_\mu \rightarrow \nu_\mu$), while the W$^\pm$ boson exchange is responsible for the β-decay, which is powering the nuclear reactions in the sun.

2.1.3 Electroweak Interaction

The weak and the electromagnetic interaction can be seen as two aspects of the same theory and that the main differences are related to the large masses of the W$^\pm$ and Z^0 bosons. Both interactions can be described within a common non-abelian gauge theory with a $SU(2)_L \times U(1)_Y$ symmetry, referred to as *electroweak* (EW) theory. Its Lagrangian density can be written as follows:

$$\mathcal{L}_{SU(2)\times U(1)} = \bar{\psi}i\gamma_\mu D^\mu \psi - \frac{1}{4} B_{\mu\nu}^i B_i^{\mu\nu} - \frac{1}{4} W_{\mu\nu}^i W_i^{\mu\nu}$$

with: $D_\mu = \partial_\mu + i\frac{g'}{2} Y B_\mu + i\frac{g}{2}\tau_a W_\mu^a$

$$B_{\mu\nu} = \partial_\mu B_\nu - \partial_\nu B_\mu$$

$$W_{\mu\nu}^i = \partial_\mu W_\nu^i - \partial_\nu W_\mu^i - g\epsilon_{ijk} W_\mu^j W_\nu^k \tag{2.3}$$

Here, ψ is the quantum field of a fermion, D_μ the covariant derivative, γ^μ are the gamma-matrices, g and g' represent the gauge coupling of the $SU(2)$ and $U(1)$ fields, and $W_{\mu\nu}$ and $B_{\mu\nu}$ the respective field strength tensors. Y is the hyper charge operator, defined as $Y = 2(Q - I_3)$ and I_3 is the third component of the weak isospin, which is a conserved quantity in electroweak interactions. The weak isospin operator is related to the Pauli matrices τ_a like $I^a = \tau_a/2$.

For left handed[2] fermions, the third component of the weak isospin is $I_3 = +1/2$ for up-types, $I_3 = -1/2$ for down-types, and $I_3 = 0$ for right handed particles. The $W^\pm$ boson couples with left handed particles exclusively. The three states of I_3 form a triplet W^i with one positive, one negative and one neutral component (W^+, W^-, W^0) for left handed particles and one electrically neutral singlet B^0 is postulated for interactions with right handed particles. The coupling strength g' of the B^0 singlet is not required to be equal to the coupling strength g of the W^i triplet. While the W^+ and W^- states correspond to the observed $W^\pm$ bosons, the neutral current mediators are linear combinations of the W^0 and B^0 eigenstates, so that the photon remains a massless particle:

$$|\gamma\rangle = sin(\theta_W)\left|W^0\right\rangle + cos(\theta_W)\left|B^0\right\rangle \tag{2.4}$$

$$\left|Z^0\right\rangle = cos(\theta_W)\left|W^0\right\rangle - sin(\theta_W)\left|B^0\right\rangle , \tag{2.5}$$

with θ_W as the electroweak mixing angle (also known as the Weinberg angle), which is related to the coupling strengths g and g' like $tan(\theta_w) = g'/g$ and has to be determined from experimental data.

2.1.3.1 Electroweak Symmetry Breaking and Mass Generation

From the pure considerations of the electroweak theory, the $W^\pm$ and the Z^0 boson are required to be massless particles, as massive gauge fields would violate the condition of local gauge invariance. However, it was shown by Higgs, Engler and Brout [6, 7] that the local gauge invariance is preserved and the gauge bosons acquire mass if they are interacting with a scalar field, called the *Higgs field* ($V(\phi)$) and its mediator, the *Higgs boson*. The Higgs field has a non-zero vacuum expectation value ν, that causes spontaneous symmetry breaking by choosing a vacuum state. As a result of the symmetry breaking, the masses of the elementary particles are generated. The coupling to the Higgs field (Yukawa coupling) for bosons reads:

$$m_W = \frac{g}{2}\nu , \qquad m_Z^0 = \frac{\nu}{2}\sqrt{g^2 + g'^2} , \tag{2.6}$$

and for fermions:

$$m_f = \frac{1}{\sqrt{2}} g_f \nu , \tag{2.7}$$

with g_f as the Yukawa coupling of a given fermion flavour.

---

[2]The helicity of a particle is referred to as left handed if the projection of the spin in the direction of motion is negative, and right handed if it is positive.

2.1.3.2   Elektroweak Currents and Unification

Any process involving only charged particles in the initial and final state can be mediated by either a Z^0 boson or a photon, with the Z^0-induced reaction being suppressed at low energy transfers, because of the particles large mass. These types of interactions are referred to as *neutral currents* (NC). In SM, flavour change of the interacting particles in NC reactions is forbidden.

Reactions that are mediated by $W^\pm$ bosons are called *charged currents* (CC) and result in the change of the flavour of one of the involved particles. For leptons, flavour change can only occur within the same generation, meaning that the lepton number of each generation ($+1$ for particles, -1 for anti-particles) is a conserved quantity. Quark wave functions are subject to the strong interaction, which is flavour blind, therefore flavour changes across generations are possible. The weak eigenstates of quarks are not equivalent to the mass-eigenstates and their *mixing* is represented by the Cabbibo-Kobayashi-Maskawa (CKM) matrix:

$$
\begin{pmatrix} d' \\ s' \\ b' \end{pmatrix} = \begin{pmatrix} V_{ud} & V_{us} & V_{ub} \\ V_{cd} & V_{cs} & V_{cb} \\ V_{td} & V_{ts} & V_{tb} \end{pmatrix} \begin{pmatrix} d \\ s \\ b \end{pmatrix}
\tag{2.8}
$$

The CKM matrix is a 3×3 unitary complex matrix, where each V_{ij} corresponds to the coupling strength of an up-type quark with a down-type quark. The values of the matrix elements can be determined experimentally [5] and are close to 1 for the diagonal elements. Off-diagonal terms, such as V_{us} or even V_{ub} contribute with small but nevertheless non-zero values, thereby accounting for cross-generation decays. Similar to the CKM matrix, the mixing of the neutrino flavour eigenstates (ν_e, ν_μ, ν_τ) with their mass eigenstates (ν_1, ν_2, ν_3) is represented by the Pontecorvo–Maki–Nakagawa–Sakata (PMNS) matrix.

Experimentally, the electroweak unification is demonstrated in the measurement of the NC and CC cross section in $e^\pm$ p scattering at the HERA collider [8]. Figure 2.2 presents the differential cross section of charged and neutral currents as a function of the momentum transfer Q^2. For collisions with low Q^2 neutral current processes with a photon as the mediator dominate over charged current processes, mediated by $W^\pm$ bosons. But with increasing Q^2, the distributions of both processes are moving closer together, until they are almost equal above $Q^2 = 10^4 \, \mathrm{GeV}^2$.

2.1.4   The Strong Interaction

The strong interaction is represented by the theory of Quantum Chromodynamics (QCD), which describes the interactions and the structure of hadrons. The strong interaction acts among quarks and is mediated by massless gluons. Quark combinations are forming hadrons: *mesons* consisting of quark and anti-quark, or *baryons* which are composed of three quarks. Experimental evidence of baryon states like

Fig. 2.2 Differential cross
sections of the charged
current (CC) and neutral
current (NC) processes as a
function of the scale Q^2. The
measurements have been
performed for e^+p and e^-p
collisions at the HERA
collider [8]

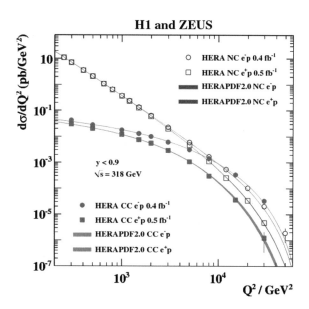

Δ^{++} required the introduction of a new degree of freedom (or quantum number), in
addition to flavour, to the wave function of a quark. Otherwise, the wave function of
the three identical quarks in Δ^{++} would violate the Pauli principle. This new degree
of freedom is called colour and can be interpreted as a charge of QCD.

The theory of QCD belongs to the groups of non-abelian gauge theories with a
$SU(3)$ symmetry [9, 10] and its Lagrangian density can be written as:

$$\mathcal{L}_{SU(3)_c} = -\frac{1}{4}F^i_{\mu\nu}F^{\mu\nu}_i + \sum_f \bar{\psi}_f(i\gamma^\mu D_\mu - m_q)\psi_f \qquad (2.9)$$

with: $\quad F^i_{\mu\nu} = \partial_\mu G^i_\nu - \partial_\nu G^i_\mu - g_s f_{ijk}G^j_\mu G^k_\mu$

$$D_\mu = \partial_\mu - i\,g_s\frac{\lambda_i}{2}G^i_\mu$$

with $F^i_{\mu\nu}$ being the field strength tensor of the gluon fields G^i_μ ($i = 1, \ldots, 8$), g_s as
the gauge coupling constant of the strong interaction and f_{ijk} the structure constants
of the Lie algebra. Note that the field strength tensor includes a self-interaction
term, missing in the EM field strength tensor. The corresponding generators λ^i are
represented by 3×3 Gell–Mann matrices which satisfy the commutations relation:

$$[\lambda^i, \lambda^j] = 2i f_{ijk}\lambda^k \qquad (2.10)$$

In the second part of the equation D_μ is the covariant derivative of the quark fields
ψ_f. The sum runs over the number of active flavours, represented by the index f.

When calculating the Feynman rules for QCD, its Lagrangian is often separated into two parts. The first part $\mathcal{L}_0$ includes all contributions from the free fields of the participating particles:

$$\mathcal{L}_0^{\text{quark}} = \bar{\psi}(i\slashed{\partial} - m)\psi \tag{2.11}$$

$$\mathcal{L}_0^{\text{gluon}} = -\frac{1}{4}(\partial_\mu G_\nu^i - \partial_\nu G_\mu^i)(\partial^\mu G^{i\nu} - \partial^\nu G^{i\mu}) - \frac{1}{2\lambda}(\partial_\mu G_i^\mu)^2 \tag{2.12}$$

$$\mathcal{L}_0^{\text{ghost}} = \partial^\mu \chi^{i*}\partial_\mu \chi^i \tag{2.13}$$

The last term of Eq. 2.12 is the so called *gauge fixing* term which is required for dealing with redundant degrees of freedom in the gluon propagator. Applying this particular method of gauge fixing introduces additional fields, so called *Faddeev-Popov ghosts* χ [11], which are needed for a complete and consistent description of the theory.

The second part, $\mathcal{L}_I$ represents the interaction terms that include the three-gluon coupling, the four-gluon coupling, the quark-gluon coupling and a gluon-ghost interaction. When expressing the scattering amplitude of a process as a Feynman diagram, these terms correspond to the vertices:

$$\mathcal{L}_I^{3\,\text{gluons}} = -\frac{g}{2}\,f_{ijk}\,(\partial_\mu G_{i\nu} - \partial_\nu G_{i\mu})\,G_j^\mu G_k^\nu \tag{2.14}$$

$$\mathcal{L}_I^{4\,\text{gluons}} = \frac{g^2}{4}\,f_{ijm}\,f_{klm}\,G_{i\nu}\,G_{j\nu}\,G_k^\mu\,G_l^\mu \tag{2.15}$$

$$\mathcal{L}_I^{\text{quark gluon}} = g\,\bar{\psi}\,\gamma^\mu\,\lambda^i\,\psi\,G_\mu^i \tag{2.16}$$

$$\mathcal{L}_I^{\text{ghost gluon}} = -g\,f_{ijk}\,(\partial_\mu \chi^{i*})\chi^j\,G^{k\mu} \tag{2.17}$$

Since gluons carry colour-charge themselves, they are self interacting, which can be seen in the Eqs. 2.14 and 2.15.

2.1.4.1 Perturbative QCD and Renormalization

The calculations of QCD observables are often computed as a perturbative expansion in the strong coupling constant α_S and the theory is therefore referred to as perturbative QCD (pQCD) [1, 9, 10]. In pQCD, any observable $\mathcal{O}$ can be approximated by a truncation of the power series, as long as α_S takes on small values.

$$\mathcal{O} = c_0 + c_1 \alpha_S + c_2^2 \alpha_S^2 + \dots \tag{2.18}$$

Similar to the fine structure constant of QED, the strong coupling is related to the gauge coupling like $\alpha_S = g_s^2/4\pi$ and the coefficients are derived from the Feynman rules. However, the calculation of contributions beyond the leading order includes loop diagrams which make the momentum of a particle within the loop go to infinity and are therefore called *ultraviolet* divergences. The responsible terms in the calculation are handled by introducing suitable counter-terms, thereby renormalizing parameters like the strong coupling constant or the quark masses to their effective values. This procedure leads to the absorption of the infinities into the renormalized parameters and introduces a renormalization scale μ_r^2 for the observables. The choice of the renormalization scale is arbitrary. However, this does not apply to the fixed order calculation, introducing significant uncertainty in the theoretical prediction for an observable at (low) fixed perturbation order.

The dependence of a renormalized parameter on μ_r is given by the Renormalization Group equations, though the parameters particular value at a scale μ_r^2 has to be determined from experimental data. For α_S, which is needed to calculate the matrix elements for the process under observation, this dependence is expressed by the $\beta(\alpha_S)$ function:

$$\beta(\alpha_S) = \mu_r^2 \frac{\partial \alpha_S}{\partial \mu_r^2} \tag{2.19}$$
$$= -\alpha_S \left(b_0 + b_1 \alpha_S + b_2 \alpha_S^2 + \mathcal{O}(\alpha_S^3) \right)$$

$$b_0 = \frac{33 - 2N_f}{12\pi}, \quad b_1 = \frac{153 - 19N_f}{24\pi^2}, \quad b_2 = \frac{77139 - 15099N_f + 325N_f^2}{3456\pi^3}$$

Here, the factor N_f corresponds to the number of quark flavours whose masses are below the scale Q. The factors b_1 and b_2 depend on the employed renormalization scheme and the ones quoted here are taken from the minimal-subtraction $\overline{MS}$ renormalization scheme [12, 13]. At leading order, the solution of Eq. 2.19 can also be written relative to the value of α_S at a scale μ_0^2:

$$\alpha_S(\mu_r^2) = \frac{\alpha_S(\mu_0^2)}{1 + b_0 ln(\mu_r^2/\mu_0^2)\alpha_S(\mu_0^2)} \tag{2.20}$$

The scale at which α_S is determined is often chosen as the mass of the Z^0 boson. The current world average for the strong coupling constant is $\alpha_S(m_Z^0) = 0.1181 \pm 0.0011$ [5]. The results of several measurements sensitive to α_S at different values of Q^2 are presented in Fig. 2.3 and include measurements from the HERA collider, the Tevatron and the CMS experiment at the LHC covering a large range in Q^2. Taking the increase of α_S towards lower values of Q^2 into consideration, Eq. 2.20 can also be written as:

$$\alpha_S(\mu_r^2) = \frac{1}{b_0 \, ln(\mu_r^2/\Lambda_{QCD}^2)} \tag{2.21}$$

where Λ_{QCD} marks the threshold under which the perturbation ansatz is no longer applicable and the theory formally diverges. For five active flavours it corresponds

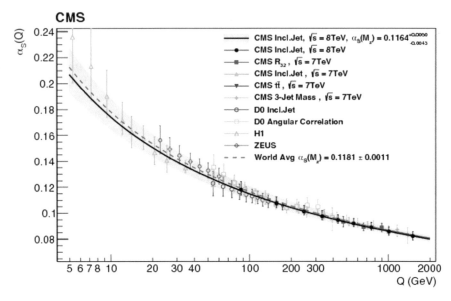

Fig. 2.3 The strong coupling constant α_S as a function of Q, as determined from the measurements by different experiments [14]

to approximately 214 MeV. Below this scale the perturbative approach breaks down because α_S becomes too large.

Due to the negative sign of the β-function (Eq. 2.19), the strong coupling constant α_S is very low at short distances or high energy scales and grows exponentially at large distances or low energy scales, which leads to an effect called *asymptotic freedom.* In contrast to the screening effect observed for the electromagnetic interaction, it is referred to as *anti-screening*, and it is the reason why quarks cannot be observed as free particles. This phenomenon is called *confinement.* If two quarks are separated by more than 10^{-15} m, the typical size of a hadron, the energy between them is sufficient to produce new hadrons. The flavour or momentum of these emerging hadrons cannot be derived from first principles but needs to be modelled phenomenologically (see Sect. 2.3.3).

Quark masses are fundamental parameters of the Standard Model. However, the bare masses appearing in the Lagrangian need to be adjusted by an infinite renormalization contribution to give a physical value and the choice of the renormalization scheme affects the value of the mass obtained. Depending on the renormalization scheme, various definitions of the quark mass exist, which require different treatments in pQCD calculations. One commonly used definition is the so called *pole mass*, which is defined as the location of the pole in the propagator [15] of the quark if it were observed as a free particle. It is well defined at each finite order of perturbation theory, though due to non-perturbative infrared effects in QCD, it retains an intrinsic uncertainty of the order of Λ_{QCD}/m_Q ($Q = $ c, b, t).

The *running mass*, defined in the $\overline{\text{MS}}$ scheme explicitly depends on the renormalization scale μ_r and provides an improved stability of the perturbative series as compared to the result in the pole mass scheme, as is demonstrated for the top quark mass in Ref. [16]. The relation between the pole mass m_Q and the $\overline{\text{MS}}$ mass $\overline{m}_Q$ is known up to four loops [17–21] and takes on the following form at one-loop order:

$$m_Q = \overline{m}_Q(\overline{m}_Q)\left[1 + \frac{4\overline{\alpha}_S(\overline{m}_Q)}{3\pi}\right] \tag{2.22}$$

As free quarks are unobservable, their masses cannot be determined via direct measurements but are derived from other mass-dependent observables. For example the charm quark mass and its scale dependence has been determined from charmed hadron production cross sections measured in $e^\pm$p-DIS experiments at HERA [22–26]. The running mass of the bottom quark has been evaluated using multi-jet events in e^+e^--collisions at LEP [27] as well as b-quark production cross sections measured in $e^\pm$p-DIS experiments at HERA [26, 28]. The recent measurements of the pole and $\overline{\text{MS}}$ mass of the top quark in proton-proton collisions are presented in Refs.[29, 30].

2.2 The Proton Structure

Protons are the lightest baryons with a mass of $m_p = 938.3$ MeV [5] and are considered stable particles. Experiments like Super Kamiokande have investigated energetically allowed decays to lighter particles [31] and conclude that the protons lifetime is more than $1.6 \cdot 10^{34}$ years for the decay channel $p \to e^+ + \pi^0$ and $7.7 \cdot 10^{33}$ years for $p \to \mu^+ + \pi^0$ (90% CL). The three valence quarks, two up and one down quark, are responsible for the quantum numbers of the proton $I(J^P) = \frac{1}{2}\left(\frac{1}{2}^+\right)$ (I = isospin, J = angular momentum, P = parity). However, the masses of the valence quarks alone ($m_u = 2.2 \pm^{+0.5}_{-0.4}$ MeV, $m_d = 4.7 \pm^{+0.5}_{-0.3}$ MeV) cannot account for the full mass of the proton and therefore its structure must be more complex.

2.2.1 Deep Inelastic Scattering

The proton structure was extensively investigated in Deep Inelastic Scattering (DIS) experiments at fixed target and at the HERA $e^\pm$p-collider, where the cross sections of NC and CC processes were measured for a large variety of final states. The Feynman diagrams of NC and CC in $e^\pm$p collisions are presented in Fig. 2.4 and their kinematics can be described by a set of variables [3]:

---

[3]The names of the four-vectors of the incoming and outgoing particles correspond to those in Fig. 2.4. $V = \gamma/Z^0, W^\pm$.

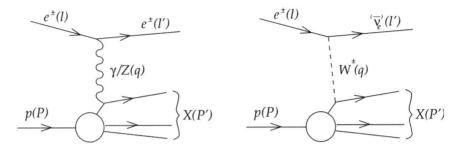

Fig. 2.4 Diagrams of neutral current (left) and charged current (right) processes in $e^\pm$ p collisions. The label X denotes the hadronic final state and the four-momenta of the incoming and outgoing particles are indicated in brackets [32]

$$Q^2 = l - l' \qquad\qquad \text{boson virtuality} \qquad (2.23)$$

$$s = (l + P)^2 \qquad\qquad e^\pm \text{ p centre-of-mass energy}^2 \qquad (2.24)$$

$$W^2 = (P + q)^2 \qquad\qquad \text{V} - \text{p centre of mass energy}^2 \qquad (2.25)$$

$$y = (P \cdot q)/(P \cdot l) \qquad\qquad \text{inelasticity} \qquad (2.26)$$

$$x = Q^2/(2P \cdot q) \qquad\qquad \text{Bjorken-}x \qquad (2.27)$$

While $\sqrt{s}$ is given by the experimental setup, the other variables need to be determined from measurements of the outgoing particles. The virtuality Q indicates the resolution capabilities of the process, meaning that with rising Q more of the proton structure can be resolved. The inelasticity y corresponds to the momentum transferred from the scattering lepton to the hadronic system in the proton rest frame, and x represents the proton's momentum fraction which is participating in the reaction.

The differential cross sections of NC (Eq. 2.28) and CC (Eq. 2.29) in DIS can be expressed in terms of three structure functions F_2, F_L and F_3, which depend on x and Q^2.

$$\frac{d^2\sigma^{NC}}{dx\,dQ^2} = \frac{2\pi\alpha_{EM}^2}{Q^4 x} \left[Y_+ F_2^{NC}(x, Q^2) - y^2 F_L^{NC}(x, Q^2) \mp Y_- F_3^{NC}(x, Q^2) \right] \quad (2.28)$$

$$\frac{d^2\sigma^{CC}}{dx\,dQ^2} = \frac{G_F^2}{4\pi x} \frac{M_W^4}{(Q^2 + M_W^2)^2} \left[Y_+ F_2^{CC}(x, Q^2) - y^2 F_L^{CC}(x, Q^2) \mp Y_- F_3^{CC}(x, Q^2) \right]$$
$$(2.29)$$

Here, $Y_\pm = 1 \pm (1 - y)^2$, and α_{EM} is the electromagnetic coupling constant. The Fermi constant G_F can be expressed as:

$$G_F = \frac{\pi\alpha_{EM}}{\sqrt{2}\, sin^2\, \theta_W^2 m_W} \, . \qquad (2.30)$$

2.2.2 Quark Parton Model

Before QCD was fully established, the simpler quark parton model (QPM) was the prevailing theory, which stated that the proton consists of charged, pointlike particles that do not interact with each other, so called *partons*. The concept of valence quarks had already been suggested independently by Gell–Mann and Zweig in 1964 [33] to explain the different hadron flavours. However, the term parton in the QPM also includes so called *sea-quark* pairs with no overall flavour.

In QPM, the proton structure function F_2 can be written as:

$$F_2 = \sum_i e_i^2 \left[xq_i(x) + x\bar{q}_i(x) \right] , \qquad (2.31)$$

with e_i as the electric charge of the parton with flavour i and q_i ($\bar{q}_i$) corresponding to a probability for a parton i to carry the x fraction of the proton momentum. The prediction was in agreement with the observations from DIS experiments, performed at SLAC [34] in the late 1960s, which measured F_2 and found that the structure function was independent of Q^2 (Bjorken scaling) [35] for $x \approx 0.3$.

Later, it was found that only about 50% of the protons total momentum could be accounted for [36]. Therefore, another type of parton, which does not interact weakly or electromagnetically, was needed to explain the missing momentum. The existence of such a particle, the gluon, was discovered at PETRA collider in 1979 at DESY [37]. Furthermore, later experiments performed at the CERN SPS or HERA collider have demonstrated the Q^2 dependence of the structure functions (scaling violation), related to the quark-gluon and gluon-gluon interactions, and to the gluon content of the proton [38].

2.2.3 Parton Distribution Functions and Factorization

In the QCD picture, the proton structure is expressed in terms of parton distribution functions (PDF) with the probability functions q_i from Eq. 2.31 modified as probabilities to find a parton i within the fraction range Δx of the proton momentum, as resolved at the scale μ^2. The PDFs are considered as universal, i.e. intrinsic property of the nucleon, being process independent. They cannot be derived from first principles but need to be determined in QCD fits to measured cross sections, taking advantage of the concept of *Factorization*. For example, the cross section for a process in pp-collisions at a scale where α_S takes on small values can be calculated using the collinear *Factorization theorem* [39–41]:

$$\sigma_{\text{pp}\to X} = \sum_{a,b} \int_0^1 \mathrm{d}x_a \mathrm{d}x_b \int f_a^{h_1}(x_a, \mu_{\text{f}}) f_b^{h_2}(x_b, \mu_{\text{f}}) \, \mathrm{d}\hat{\sigma}_{ab\to n}(\mu_{\text{f}}, \mu_{\text{r}}) \tag{2.32}$$

$$= \sum_{a,b} \int_0^1 \mathrm{d}x_a \mathrm{d}x_b \int \mathrm{d}\Phi_n \, f_a^{h_1}(x_a, \mu_{\text{f}}) f_b^{h_2}(x_b, \mu_{\text{f}}) \frac{1}{2\hat{s}} |\mathcal{M}_{ab\to n}|^2(\Phi, \mu_{\text{f}}, \mu_{\text{r}})$$

$$\tag{2.33}$$

Here, $x_{a,b}$ are the momentum fractions carried by the partons a and b, whereas $f_{i/j}(x_{a,b}, \mu_F^2)$ are the PDFs of the partons concerned in the production process and $\hat{\sigma}_{ab\to n}(\mu_{\text{f}}, \mu_{\text{r}})$ represents the partonic cross section of the process $a + b \to n$. The latter is a combination of the matrix element $\mathcal{M}$, describing the interactions between the particles, the particle flux $1/2\hat{s} = 1/(x_a x_b s)$ with s as the centre of mass energy, and an integration over the phase space $\mathrm{d}\Phi$ of the final state particles n.

Factorization requires the introduction of a new scale, the so called *factorization scale* μ_{f}, which is needed to handle infrared divergences, arising from this approach. Infrared divergences occur, when final state quarks radiate soft gluons or a high momentum gluon is emitted collinear with a low momentum quark. Therefore, factorization can also be described as the separation of the *hard* and *soft* regime of a collision. In many calculations the factorization scale μ_{f} it is set equal to the renormalization scale μ_{r}, which is typically chosen close to the energy involved in the hard scattering process. For DIS experiments this corresponds to the virtuality Q^2, whereas processes that involve the production of heavy particles (i.e. $W^\pm/Z^0$ bosons, heavy quarks), it is often set to the mass of the heaviest particle or the transverse mass $M_T = \sqrt{m^2 + p_T^2}$.

2.2.4 Evolution of PDFs

The dependence of the PDFs on the factorization scale μ_{f} can be derived from the renormalization group equations and expressed as so called PDF evolution formalisms which differ in the ordering of their parton emissions. The Dokshitzer-Gribov-Lipatov-Altarelli-Parisi (DGLAP) evolution equations [42–47], belong to the collinear formalisms and the emissions are ordered by their longitudinal momenta. In the DGLAP equations, the quark and gluon distributions are coupled:

$$\frac{\partial q(x, \mu_{\text{f}}^2)}{\partial ln(\mu_{\text{f}}^2)} \propto \int_x^1 \frac{\mathrm{d}z}{z} \left[q(z, \mu_{\text{f}}^2) P_{qq}\left(\frac{x}{z}\right) + g(z, \mu_{\text{f}}^2) P_{qg}\left(\frac{x}{z}\right) \right] \tag{2.34}$$

$$\frac{\partial g(x, \mu_{\text{f}}^2)}{\partial ln(\mu_{\text{f}}^2)} \propto \int_x^1 \frac{\mathrm{d}z}{z} \left[q(z, \mu_{\text{f}}^2) P_{gq}\left(\frac{x}{z}\right) + g(z, \mu_{\text{f}}^2) P_{gg}\left(\frac{x}{z}\right) \right] \tag{2.35}$$

Here, P_{ij} (i,j = q,g) are the splitting functions, which represent the probability that a parton i radiates parton j, with j carrying the fraction z of i's original momentum. The corresponding Feynman diagrams of the leading order splitting functions are

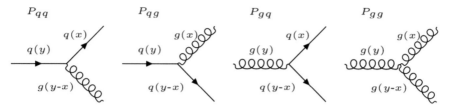

Fig. 2.5 Illustration of the leading order splitting functions for the DGLAP evolution equations [48]

presented in Fig. 2.5. These can be calculated as a perturbative expansion of α_S and take on the following form at leading order:

$$P_{gg}(z) = 6 \left(\frac{z}{(1-z)_+} + \frac{1-z}{z} + z(1-z) + \left(\frac{11}{12} - \frac{N_f}{18} \right) \right) + \frac{1}{2} \beta_0 \delta(1-z)$$

(2.36)

$$P_{qq}(z) = \frac{4}{3} \left(\frac{1-z^2}{(1-z)_+} + \frac{3}{2} \delta(1-z) \right)$$

(2.37)

$$P_{qg}(z) = \frac{1}{2} \left(z^2 + (1-z)^2 \right)$$

(2.38)

$$P_{gq}(z) = \frac{4}{3} \left(\frac{1 + (1-z)^2}{z} \right) ,$$

(2.39)

with $N_c = 3$ corresponding to the number of colours, $\beta_0 = 11 - 2/3 N_f$ and N_f as the number of quark flavours. Using the DGLAP equations, it is possible to evolve the PDFs from an initial scale μ_0^2 to any (high enough) values of μ^2. The scheme is applicable in the medium to high x region. However, the x dependence of the PDFs cannot be calculated from first principles and therefore has to be extracted from experimental data. This is performed by confronting the measurements of the proton structure functions or of the cross sections of NC and CC to the theoretical predictions, using the factorization theorem. At the starting scale μ_0^2, an initial shape of the PDFs as a function of x is assumed and the DGLAP equations are used for the μ^2 evolution. As additional constraints on the x-dependence of the PDFs, the *QCD sum rules* are applied:

The *momentum sum rule* states that the sum of all momentum fractions carried by the different quark flavours f_i and gluons g constitutes the total momentum of the proton:

$$\sum_i \int_0^1 dx \, x f_i(x) + \int_0^1 dx \, x g(x) = 1$$

(2.40)

Two additional sum rules can be derived by considering the valence quark content of the proton. As it contains two up and one down quark, the net content of both particles with regard to their anti-particles is given by:

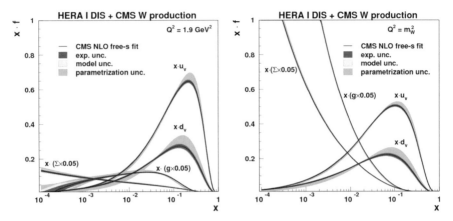

Fig. 2.6 Parton distributions functions, obtained at NLO for the scales of $\mu_0^2 = 1.9\,\text{GeV}^2$ (left) and $\mu^2 = m_W^2$ [49]. The different contributions to the PDF uncertainties are indicated by bands of different colours

$$\int_0^1 dx u_v(x, \mu^2) = \int_0^1 dx \left[u(x, \mu^2) - \bar{u}(x, \mu^2) \right] = 2 \qquad (2.41)$$

$$\int_0^1 dx d_v(x, \mu^2) = \int_0^1 dx \left[d(x, \mu^2) - \bar{d}(x, \mu^2) \right] = 1 \qquad (2.42)$$

An example of the proton's PDFs is shown in Fig. 2.6, where the valence quark, sea quark and gluon distributions are presented as a functions of x, evaluated at the scales $\mu^2 = 1.9\,\text{GeV}^2$ and $\mu^2 = m_W^2$. The gluon and sea quark distributions are scaled down by a factor of 20 for demonstration purposes.

Other PDF evolution formalisms like the Balitsky-Fadin-Kuraev-Lipatov (BFKL) [50, 51] are more suited to handle the evolution of PDFs in the low-x region. The BFKL scheme uses so called unintegrated PDFs (uPDF) to evolve PDFs from an initial value x_0 to lower values at a fixed scale μ_0^2, usually chosen as the hard scattering scale. The scheme is using the longitudinal momentum fraction x to order the emissions and is explicitly depending on the transverse momentum of the partons. The CCFM [52–55] evolution formalism aims to combine the DGLAP and BFKL approaches and applies a strong angular ordering of the emissions. Using this scheme also requires the use of transverse momentum dependent PDFs.

2.2.5 Strange Quark Content of the Proton

The strange quark distribution is the least constrained PDF among the light quarks. For measurements like the determination of the $W^\pm$ boson mass at hadron colliders [56], uncertainties from PDFs, especially the strange quark distribution, are still

Fig. 2.7 Feynman diagram of a ν_μ induced charged current charm di-muon production [59]

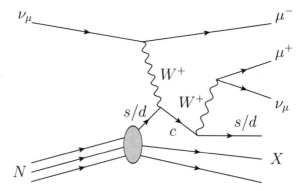

a dominant source of uncertainty. Results constraining the s-quark distribution stem, for a large part, from the analysis of charm production in charged current deep-inelastic (anti-)neutrino-nucleus scattering. In these measurements, the cross section of oppositely-charged di-muon production in (anti-)neutrino-nucleon DIS was determined by the CCFR, NuTeV and NOMAD collaborations [57–59]. The Feynman diagram of the investigated process is presented in Fig. 2.7. The muon carrying the higher p_T is considered to originate from the neutrino scattering, whereas the oppositely charged muon is assumed to be a decay product of the charmed particle (i.e. D^*, $D^\pm$, Λ_c).

Additionally, the cross section of inclusive charm production from neutrino and anti-neutrino charged-current in nuclear emulsions was measured by the CHORUS collaboration [60]. In this experiment it was possible to separate the contributions from different charmed particles by measuring the flight length and the momentum of the hadrons, thereby extracting the charmed particle production fractions. The cross sections extracted by CHORUS are complementary to the aforementioned experiments, as they do not depend on the branching ratio $\mathcal{B}(c \to \mu)$.

All of the ν-scattering measurements have a high sensitivity to the strange quark content of the nucleon sea, since the contribution from d-quarks to charm production is Cabibbo suppressed. PDFs extracted from ν-DIS experiments have to take the nuclear corrections of the target into account, which introduces additional sources of uncertainties. However, using the ratio of charm-dimuon to the full CC production cross section, which is provided by both NOMAD and CHORUS, largely cancels nuclear corrections in the extracted PDFs.

The strange quark distribution is often presented relative to the PDFs of the other light sea quarks via the so called *strangeness suppression* r_s:

$$r_s(x, \mu^2) = \frac{s(x, \mu^2) + \bar{s}(x, \mu^2)}{\bar{u}(x, \mu^2) + \bar{d}(x, \mu^2)} . \tag{2.43}$$

The corresponding strangeness suppression factor κ is the integral of r_s over the full x-range, evaluated at a chosen scale μ^2. It has been determined by the NOMAD collaboration at the scale of $\mu^2 = 20\,\mathrm{GeV}^2$ as $\kappa = 0.591 \pm 0.019$.

The data of all three experiments is used in various QCD fits to determine the strange quark content of the proton. Due to the much larger statistics of the NOMAD measurements, the uncertainty in the strange quark distribution extracted is a factor 2 lower than for analyses using the CCFR or NuTeV data alone [59]. The impact of the NOMAD and CHORUS results on the strange quark distribution has been investigated [61] using the ABM12 PDF set [62] as a baseline, which already includes the CCFR and NuTeV data. The NOMAD and CHORUS datasets are added to the fit individually, as well as together, and it was found that adding only the NOMAD data pulls the strange quark distribution somewhat down, whereas using the CHORUS data has the opposite effect. Including both datasets simultaneously results in a strange sea distribution that is in good agreement with the baseline of using the CCFR and NuTeV data alone, as both dataset compensate the pulls on the distribution. Among the currently available PDF sets, only the ABMP16 [63] PDF sets include the latest NOMAD results and therefore show the most precise strange quark distribution. The CT14nlo [64] PDF set only includes the NuTeV and CCFR data, whereas MMHT14nlo [65] additionally includes the CHORUS measurements. The NNPDF3.1nlo [66] set uses the NuTeV and CHORUS results, as well as the 7 TeV measurement of W + c, performed by the CMS collaboration [67]. The strange quark distribution and strangeness suppression at the scale of $\mu_{\mathrm{f}}^2 = m_{\mathrm{W}}^2$ of the global PDF sets ABMP16nlo, CT14nlo, MMHT14nlo and NNPDF3.1nlo are presented in Fig. 2.8 and are in good agreement with each other. In addition, the distributions from the ATLASepWZ16nnlo analysis are also shown in Fig. 2.8, and do not agree with the global PDFs, showing an unsuppressed strange quark distribution. The details are discussed in the following.

At hadron colliders, such as Tevatron or the LHC, the inclusive production of electroweak bosons (Drell–Yan processes, DY) provides constrains for all types of light quarks and is therefore indirectly sensitive to the strange quark distribution in the proton. The total cross sections of $W^\pm$ and Z^0 receive contributions from strange quarks through processes like $u + \bar{s} \rightarrow W^+$, $\bar{u} + s \rightarrow W^-$ and $s + \bar{s} \rightarrow Z^0$. The contributions of different quark-quark hard scattering processes in $W^\pm$ and Z^0 production in hadron collisions are illustrated in Fig. 2.9. The differential cross sections of $W^\pm$ and Z^0 production can be determined with good accuracy and theoretical predictions are available at NNLO precision in programs like DYNNLO [68, 69] or FEWZ [70].

The ATLASepWZ12 [72] and ATLASepWZ16 [73] analyses explore the indirect sensitivity of the inclusive $W^\pm$ and Z^0 production to the strange quark content of the proton, in combination with the $e^\pm p$ scattering cross-sections from HERA. Predictions for these measurements are available at next-to-next-to-leading order (NNLO) and the ATLAS results of inclusive $W^\pm$ and Z^0 production are measured at 7 TeV with a precision of 1.8%. It was observed that the strange quark distribution extracted in the ATLAS analyses [72, 73] is unsuppressed, indicating an almost symmetric flavour composition of the light quark sea in the kinematic range of $x \geq 0.01$.

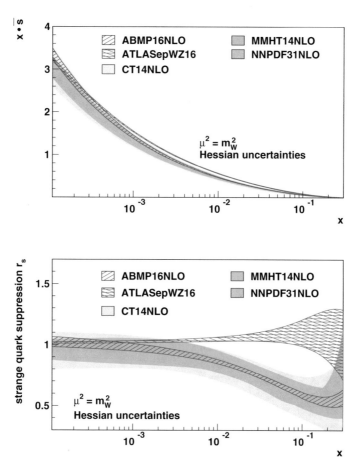

Fig. 2.8 The strange quark distribution (upper) and strangeness suppression (lower) as a function of x obtained at the scale $\mu_f^2 = m_W^2$ for the ABMP16nlo, ATLASepWZ16nnlo, CT14nlo, MMHT14nlo and NNPDF3.1nlo PDF sets. The PDF uncertainties, resulting from the experimental uncertainties of the input datasets are evaluated using the Hessian method

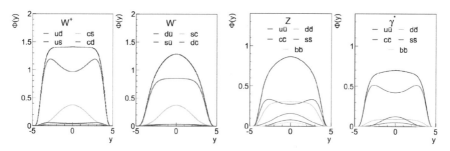

Fig. 2.9 LO quark-quark processes (lines of different colour) contributing to the $W^\pm$ and Z^0/γ^* cross sections (black line) as a function of the boson rapidity [71]

The ATLASepWZWjet19nnlo analysis [74] also includes the measurement of W-boson production in association with at least one jet. Due to the requirement of at least one jet to be present in an event, this measurement probes the sea quark distributions at higher values of x, compared to the inclusive W production analysis. The resulting strange quark distribution and strangeness suppression factor show a suppression for $x \geq 0.23$ and significantly reduced tension with global PDFs.

The possible reasons for the enhanced strange quark distribution observed by ATLAS have been investigated in Refs. [61, 75]. The influence of the parametrization and used datasets on the strangeness suppression and isospin asymmetry have been studied separately to determine the source of the deviation in the strange quark distribution with regard to other, global PDF sets. All QCD analyses of this study have been performed at NNLO accuracy, using the ABMP16 setup with a starting scale of $\mu = 3\,\mathrm{GeV}$ and the $n_f = 3$ flavour scheme. The ABMP16 parametrization has a total of 25 free parameters, whereas the parametrization employed by ATLAS has 15 free parameters, and is based on the HERAPDF [8] QCD analyses, with the restrictions of $A_{\bar{u}} = A_{\bar{d}}$ and $B_{\bar{u}} = B_{\bar{d}} = B_{\bar{s}}$. Both parametrizations are used in fits to the same datasets, based on a selection of the ABMP16 PDF set, though the DY datasets from the LHC and Tevatron were replaced with deuteron DIS data. It has been shown that the replacement datasets are in agreement with the $W^{\pm}$ and Z^0 boson collider measurements after the deuteron corrections are taken into account, and provide constrains on light quark PDFs in the high-x region, where the enhancement is observed.

Figure 1 of Ref. [75] presents the strangeness suppression and isospin asymmetry of the ABMP16 PDF set, the results of the fit using the ABMP16 parametrization, and the results of the fit using the ATLAS parametrization. The strangeness suppression of the fit performed with the ABMP parametrization is in good agreement with the nominal ABMP16 distribution in the region of $x \geq 0.01$, but deviates towards lower values of x due to the missing constrains provided by DY collider data. The QCD analysis performed with the ATLAS parametrization shows an enhancement of r_s in the region $x \geq 0.01$. When comparing the isospin asymmetry of the different analyses though, it is observed that the ATLAS distribution is significantly lower than the ones obtained with the ABM parametrizations in the region of $0.02 \leq x \leq 0.1$, meaning that the observed strangeness enhancement is compensated by a suppression of the down quark sea.

In turn, the influence of the datasets used in ATLASepWZ16nnlo analysis was investigated by fitting the same datasets, but using the ABMP16 parametrization. An enhancement at large x, with regard to the nominal ABMP16 PDF set is observed, through this effect is largely mitigated when the fixed target DY data, measured by the E866 [76] experiment is included in the fit. The study [75] concludes that the strangeness enhancement observed in the ATLASepWZ12nnlo and ATLASepWZ16nnlo PDF sets is due to the limited flexibility of the chosen parametrization.

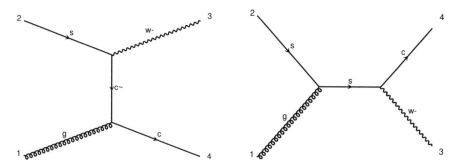

Fig. 2.10 Feynman diagrams of W + c production at the LHC at LO in pQCD. Diagrams made by MADGRAPH5_aMC@NLO [81]

2.2.5.1 Probing the Strange Quark Distribution in Proton-Proton Collisions: W + charm

The production of a W boson associated with a single charm quark has been suggested as a suitable process to study the strange quark distribution at hadron colliders early on [77–80]. This process probes the strange quark content of the proton, as W + c is dominantly produced by the hard scattering of a strange quark and a gluon at leading order $\bar{s} + g \rightarrow W^- + c$ and $s + g \rightarrow W^+ + \bar{c}$. The two Feynman diagrams contributing to the cross section at leading order are presented in Fig. 2.10. Like in the case of ν-scattering experiments, contributions from d quarks are expected as well, though at a much smaller rate. In all diagrams presented in this section, the strange quark can be substituted for a d-quark and due to the presence of a d-valence quark in the proton, a small asymmetry between the $W^+ + \bar{c}$ and $W^- + c$ cross sections is expected.

The first measurements of W + c were performed at the Tevatron $p\bar{p}$-collider at a centre-of-mass energy of $\sqrt{s} = 1.96$ TeV, by both the CDF [82, 83] and D0 [84] collaborations. The $W^\pm$ bosons and charm quarks were reconstructed via their leptonic decays, resulting in an event signature with two oppositely charged muons, of which the one originating from the charm decay is contained in a jet. In these measurements, the inclusive cross section of W + c, as well as the ratio of $\sigma(W + c)/\sigma(W^\pm + 1\,\text{jet})$ has been determined with an uncertainty of $20 - 30\%$, which is dominated by the statistical uncertainty.

At the LHC, the measurement of W + c was first performed by the CMS collaboration at a centre-of-mass energy of $\sqrt{s} = 7$ TeV [67]. The analysis strategy for the measurement is similar to the ones employed at the Tevatron, though additional methods of charm tagging were utilized. These included the reconstruction of charmed mesons contained in a jet via the hadronic decays of $D^\pm \rightarrow K^\mp \pi^\pm \pi^\pm$ and $D^*(2010)^\pm \rightarrow D^0 \pi^\pm$ with $D^0 \rightarrow K^\mp \pi^\pm$. The fiducial W + c cross section was measured inclusively, and as a function of the pseudorapidity of the lepton, originating from the $W^\pm$ boson decay, which is presented in Fig. 2.11.

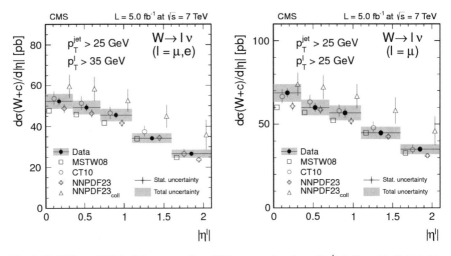

Fig. 2.11 Differential fiducial cross section of W + c as a function of $|\eta^l|$ (left) and $|\eta^\mu|$ (right) in $\sqrt{s} = 7$ TeV pp-collisions, determined by CMS [67]. The measured values are presented as filled circles, with black error bars (shaded bands) indicating the statistical (total) uncertainty of the measurement. The results are compared to predictions calculated with MCFM [86–88] in combination with different PDF sets, presented as symbols of different styles and colors

The ATLAS collaboration also determined the W + c-jet cross section at $\sqrt{s} = 7$ TeV [85] as a function of the lepton pseudorapidity. In this analysis, the reconstruction of $D^*(2010)^\pm$ mesons also included the D^0 decays $D^0 \to K^\mp \pi^\pm \pi^\mp \pi^\pm$ and $D^0 \to K^\mp \pi^\pm \pi^0$. In addition to the W + c-jet cross section, the cross section of $\sigma(W + D^*(2010)^\pm)$ and $\sigma(W^\mp + D^\pm)$ was determined. Though these measurements are quite precise, they are performed at particle level only, and are therefore not used in any global QCD analysis.

An NLO QCD analysis, including the CMS results of W + c at 7 TeV, was performed to assess the constraining power of that measurement on the s-quark distribution of the proton [49]. There, the results of [67] were used in a QCD analysis together with the measurements of NC and CC DIS cross section at HERA [89] and the CMS results for the lepton charge asymmetry in $W^\pm$ boson production at 7 TeV. The extracted strange quark distribution and strangeness suppression at the scales $\mu^2 = 1.9\,\text{GeV}^2$ and $\mu^2 = m_W^2$ are presented in Fig. 2.12. The strangeness suppression factor of this QCD analysis has been determined at the same scale as the one published by the NOMAD collaboration ($\mu^2 = 20\,\text{GeV}^2$), and a value of $\kappa = 0.52^{+0.12}_{-0.10}(\text{exp.})\,{}^{+0.05}_{-0.06}(\text{model})\,{}^{+0.13}_{-0.10}(\text{param.})$ was extracted. Both results are compatible within uncertainties.

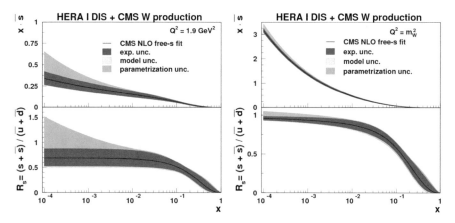

Fig. 2.12 The strange quark distribution (upper panel) and strangeness suppression (lower panel) as a function of x, obtained at the starting scale $\mu_f^2 = 1.9\,\text{GeV}^2$ (left) and $\mu_f^2 = m_W^2$ (right). The full band represents the total uncertainty, while the contributions from different uncertainties are represented by the bands of different colours [49]

2.3 Theoretical Predictions

In the following, the calculation of fixed order predictions and Monte Carlo event simulations are described. Full event simulations include matrix element calculations, determined at a particular order of pQCD and approximations of non-perturbative effects that require phenomenological modelling, as well as a simulation of the detector response. They are used for comparisons with experimental data or to simulate the performance of future experiments. However, since not all functionalities are available for every calculation or simulation, dedicated programs exist for each of the different steps and can be interfaced with others. Figure 2.13 illustrates the different steps in sim ulating proton-proton collisions.

2.3.1 Matrix Element Calculations: MADGRAPH, POWHEG, MCFM

The parton level cross section of a process of interest (i.e. W + c) is determined using the factorization theorem (see Sect. 2.2.3) which involves the calculation of the matrix elements $\mathcal{M}_{ab\to n}$ contributing at fixed orders of pQCD. The probability of two partons interacting and their momentum in the initial hard scattering are defined by the PDF set used in the simulation. The calculation of leading order matrix elements is largely automated at this point and has been implemented in several so called *event generators*, such as PYTHIA [91] or MADGRAPH [92].

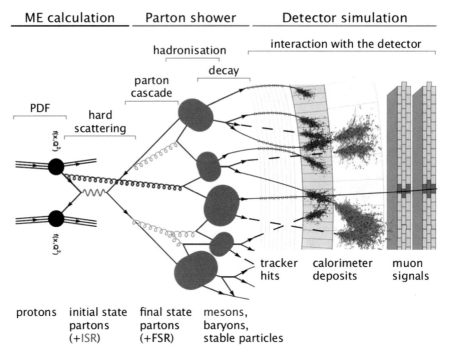

ME calculation ___ Parton shower ___ Detector simulation

Fig. 2.13 Illustration of the different steps involved in the simulation of a full event, using Monte Carlo methods [90]

Due to the large scale dependence of the LO calculations, NLO or even NNLO generators are preferred, though the divergences arising from real and virtual corrections make calculations beyond LO very complex. The strategies developed to deal with this issue can be categorized into two general approaches [93], the so called *phase space slicing* method [94, 95] and the *infrared subtraction* method [96–102], with the latter being the preferred tool in many current event generators. The subtraction method is based on the observation that the soft and collinear divergences in the real corrections $\mathcal{R}$ exhibit a universal structure, which can be calculated using a convolution of LO (Born-level $\mathcal{B}$) matrix elements and splitting kernels $\mathcal{S}$. The modified term for the real emissions $[\mathcal{R} - \mathcal{B} \otimes \mathcal{S}]$ is infrared finite and can therefore be integrated over the full phase space $\tilde{\Phi}_{n+1}$. Furthermore, the divergences arising in the calculation of the virtual corrections $\mathcal{V}$ possess the same structure as the ones for the real emission corrections, but with an opposite sign. Therefore, the same subtraction terms can be added to the virtual corrections, ensuring the cancellation with the real part of the calculation.

$$\sigma^{\mathrm{NLO}} = \int_n \mathrm{d}\Phi_n^{(4)} \mathcal{B} \ + \alpha_{\mathrm{S}} \int_{n+1} \mathrm{d}\Phi_{n+1}^{(4)} \left[\mathcal{R} - \mathcal{B} \otimes \mathcal{S} \right]$$

$$+ \alpha_{\mathrm{S}} \int_n \mathrm{d}\Phi_n^{(D)} \left[\mathcal{V} + \mathcal{B} \otimes \int_1 \mathrm{d}\tilde{\Phi}_1^{(D)} \mathcal{S} \right] \quad (2.44)$$

Though is was originally only available for massless partons, the method has been adapted to include massive fermions [97, 103] in the calculations.

Among the NLO event generators, MADGRAPH5_aMC@NLO [81] is capable of computing tree-level and one-loop amplitudes for arbitrary processes at LO and NLO accuracy, based on the Frixione, Kunzst and Signer (FKS) subtraction method [98, 99]. Moreover, the generated events can include higher order corrections of a given process in the form that additional real emissions are considered in the calculations. Another commonly used NLO event generator using the FKS subtraction formalism is POWHEG [104–106], which provides calculations of heavy flavour productions in hadron collisions at NLO accuracy. Both generators are widely used as the starting point of full event simulations as they can be interfaced to other programs for parton shower and hadronization and provide suitable matching and merging procedures for such cases.

Alternately, Monte Carlo for FeMtobarn processes (MCFM) [86–88] uses the subtraction formalism of Catani and Seymour [96] to calculate parton level cross sections. MCFM is specialized in calculating production processes that include $W^{\pm}$, Z^0 and H bosons, as well as heavy quarks (c, b, t) in the final state. Most of the implemented processes are available at NLO and include the spin correlations in the decays, though several processes are also available at NNLO. In MCFM, the NLO corrections to W + c are calculated following the same strategy as was used for the associated production of a W boson and a top quark (W + t) [107]. The massive quarks in the final state are handled by using an extension of the CS dipole formalism [97]. The mass of the heavy quark is treated in the pole mass scheme, whereas the strong coupling constant at the renormalization scale is evaluated according to the $\overline{\mathrm{MS}}$ scheme.

2.3.1.1 Fixed Order Calculations of W+charm Production

The first calculation of the NLO QCD corrections [108, 109] is based on the phase space slicing method [94, 95] developed for the calculation of jet cross sections with an arbitrary number of partons in the final state. It has been adapted to include massive quarks by using the Collins–Wilczek–Zee [110] scheme, where the heavy quarks decouple for small momenta. In these calculations, charm quarks are not considered as an active flavour of the proton, therefore diagrams with charm quarks in the initial state are not included.

Singularities arising in the calculation are isolated by slicing up the phase space and separating it into a hard region, containing no singularities, and a region in which the final state parton is either soft or emitted collinear with one of the initial

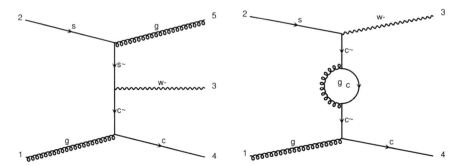

Fig. 2.14 Feynman diagrams of real (left) and virtual (right) loop corrections to the $g + s \rightarrow W + c$ production process at next-to-leading order in pQCD

state partons. Cases where the final state heavy quark is radiating a gluon exhibit no collinear singularities because they are shielded by the presence of the heavy quark mass.

The hard region is constructed such that all partons are resolved as separate jets. A small, but otherwise arbitrary theoretical cutoff value s_{min} is defined and the invariant mass calculated from the four-momenta of two neighboring partons $s_{ij} = 2 P_i \cdot P_j$ must fulfill $s_{ij} \geq s_{min}$. Here, it is possible to work in 4 dimensions and perform the integration numerically. In the soft and collinear region, one or two of the s_{ij} fall below the cutoff threshold s_{min}. The integration is performed analytically in d dimensions. Tests of the calculations where s_{min} was varied, show that the choice of the value has a strong influence on the individual contributions from the hard and the soft- and collinear regions, but almost no influence on the size of the cross section, as long as s_{min} is set to reasonably small values.

Any singularities in the soft and collinear region, not covered by the renormalization of the quark mass, factorize into a universal factor that is multiplied with the LO diagrams:

$$\alpha_S \frac{N_c}{8\pi} P_{qg \rightarrow q}(z) \, ln \left(\frac{M^2}{m^2} \right) \tag{2.45}$$

Here, M is the upper limit of the heavy quark (pair) invariant mass defining the collinear region, m is the heavy quark mass, N_c is the number of colour states and $P_{qg \rightarrow q}(z)$ is the splitting function:

$$P_{qg \rightarrow q}(z) = \lim_{\delta \rightarrow 0} 2 \left(1 - \frac{1}{N_c^2} \right) \left(\left(\frac{1+z^2}{1-z} \right) \theta(1 - z - \delta) + \left(\frac{3}{2} + 2ln(\delta) \right) \delta(1 - z) \right) \tag{2.46}$$

Examples for real and virtual one-loop contributions to the LO process of $s + g \rightarrow W + c$ are presented in Fig. 2.14. Additional NLO tree diagrams for $2 \rightarrow 3$ processes involving $g + g \rightarrow s + W + c$ and $q + s \rightarrow q + W + c$ are shown in Fig. 2.15.

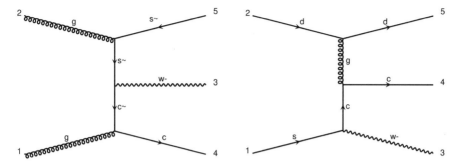

Fig. 2.15 Feynman diagrams of W + c production at the LHC, not involving $g + s \rightarrow W + c$ at next-to-leading order in pQCD. All diagrams made by MADGRAPH5_aMC@NLO [81]

2.3.2 Parton Shower Description

As higher orders, matrix element calculations become more and more difficult and computationally intensive, radiative corrections to LO or NLO calculations are often approximated using a *parton shower* algorithm. There, additional radiations of the types $q \rightarrow qg$, $q \rightarrow q\gamma$, $g \rightarrow gg$, $g \rightarrow q\bar{q}$ and $l \rightarrow l\gamma$, $\gamma \rightarrow l\bar{l}$ are simulated by evaluating the probability of an emission occurring between two evolution scales μ_1 and μ_2, with $\mu_1 > \mu_2$. The evolution of a parton shower is considered universal, and is therefore independent from the production process of the showered partons (e^+e^-, $e^{\pm}p$, etc.).

To avoid both, soft and collinear divergences when using a parton shower, only *resolvable* emissions above a cutoff scale μ_0 are permitted, thereby ensuring a finite emission probability. This scale is often chosen as $\mu_0 = 1\,\text{GeV}$ in order to remain in the region of phase space where pQCD is applicable. The inverse, meaning the probability of no resolvable emission occurring between two scales is given by the Sudakov form factor [93]:

$$\Delta_i(\mu_1^2\,,\,\mu_2^2) = exp\left(\int_{\mu_2^2}^{\mu_1^2} \frac{\mathrm{d}k^2}{k^2} \frac{\alpha_S}{2\pi} \int_{\mu_0^2/k^2}^{1-\mu_0^2/k^2} \mathrm{d}z\, P_{ij}(z)\right)\,, \qquad (2.47)$$

with $P_{ij}(z)$ representing splitting functions.

Two commonly used parton showers are implemented in PYTHIA [91] and HERWIG++ [111] and can be interfaced with other generators such as MADGRAPH or POWHEG for the initial calculations of the matrix elements. PYTHIA uses a so called *transverse-momentum-ordered* parton shower with the evolution scale corresponding to the transverse momentum of the radiated parton $\mu = p_T$, whereas HERWIG uses a *angular-ordered* scheme with $\mu \approx E^2\theta^2$. In the latter case, E corresponds to the energy of the radiated parton and θ representing the angle between the participating partons.

The evolution of a parton shower starts at a starting scale μ_{max}, typically chosen as the scale of the hard scattering process. New emissions are generated according to the solution of $\Delta_i(\mu_1^2, \mu_2^2) = \rho$, where ρ is a randomly chosen number between 0 and 1. If the resulting new scale μ_2 is above μ_0, a new parton at the scale of μ_2 is generated and z is chosen according to P_{ij}. This procedure is iterated for all partons in an event, until the cutoff scale μ_0 is reached and the algorithm is terminated.

2.3.2.1 Matching and Merging in MC

The combination of ME calculations with a parton shower cannot be accomplished by independently applying a PS algorithm to the final state partons of ME calculations. Doing so introduces double counting of contributions that involve additional emissions to the hard scattering. These can be generated by a ME calculation involving $n + 1$ partons in the final state, as well as a calculation with n final state partons, where the hard radiation originates from the parton shower. Figure 2.16 illustrates possible scenarios that lead to double counting when ME calculations and PS are used independently. These issues are resolved by applying *matching* or *merging* procedures, though the methods require different approaches for LO and NLO calculations due to the radiative corrections already included in NLO calculations.

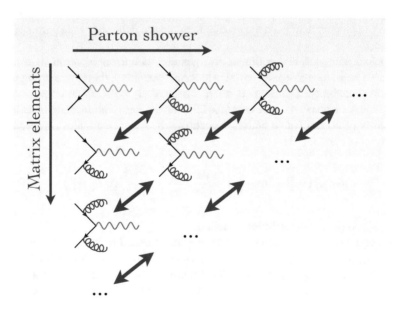

Fig. 2.16 Illustration of double counting in the combined use of matrix element (ME) calculations and parton shower (PS) algorithms. Real emissions originating from ME calculations are marked in red, whereas contributions from the PS are shown in blue [112]

Generally, parton showers are well suited to describe the soft and collinear region, but fail to give an accurate description of hard radiation, whereas LO ME calculations provide a good description of hard, well separated partons in the final state but do not work well in the soft an collinear limit. For LO matrix elements, this is resolved by introducing a cutoff scale Q_c. The approach avoids double counting by only allowing ME contributions if they are above Q_c and PS contributions if they are below Q_c, thereby *merging* the ME and PS distributions. The choice of Q_c is arbitrary and introduces bias if chosen too high or too low. It is therefore important to assure that the ME and PS distributions have a similar shape close to Q_c.

When dealing with NLO matrix elements, another source of double counting is introduced, as the virtual corrections from the NLO calculations produce the same final states as the non-emission probability of the Sudakov suppression in the PS. The MC@NLO [113] method removes the double counting with the parton shower by subtracting the contributions from the parton shower in the NLO calculations. However, this method can result in negative event weights and therefore a larger sample is needed to reach the same statistical accuracy as other generators that only have positive weights. An alternative approach, implemented in POWHEG (v2.0) [114], requires that the first emission must be the hardest in scale, whereas all subsequent emissions must be below the initial scale of the radiated parton. Therefore, the parton shower can only generate softer emissions, as any that are above the initial scale are vetoed. The hardness of the first emission depends on a dampening parameter h_{damp}.

The FxFx [115] technique, used in MADGRAPH5_aMC@NLO [81], is combining several NLO calculations for $X + 0, 1, \ldots, n$-jets. The resulting matrix elements are multiplied with the appropriate Sudakov factors, and a merging scale Q_c is defined for the contributions from the ME and the PS.

2.3.3 *Fragmentation and Hadronization*

The transition from a partonic final state to a hadronic final state cannot be calculated in pQCD and therefore relies on phenomenological models that are tuned to match the observations in fixed target and collider experiments. The fragmentation probability of a quark with flavour f hadronizing into a specific particle (e.g. $c \rightarrow$ D*$(2010)^{\pm} = 2429 \pm 0.0049$ [116]) is considered a universal property and therefore independent of the production mechanism of the quark. This property has been tested by comparing the light quark production in e^+e^-, $p\bar{p}$, γp and $\gamma\gamma$ collisions and so far no significant deviation from the presumed hadron production universality has been observed [117].

In hadronization models, the fraction of the quark's longitudinal momentum transferred to the emerging hadron (z) is described by the so called *fragmentation function*. The Lund [118] parametrization, used in the *Lund String-Model* [118, 119], which is implemented in PYTHIA takes on the following form:

$$f(z) = \frac{1}{z}(1-z)^a \, exp\left(\frac{-b\,m_t^2}{z}\right), \tag{2.48}$$

It is multiplied with the Bowler modification $1/z^{r_q\,b\,m_q^2}$ [120] for the treatment of heavy quarks. Here, a and b are free parameters $m_t = \sqrt{m_q^2 + p_T}$ corresponds to the transverse mass of the quark and r_q depends on the quark flavour and used generator. The string model assumes a strong field between two quarks that are drifting apart. The energy density of the string connecting both particles is of the order of 1 GeV/fm and will be picked up by short range $q\bar{q}$ fluctuations of the vacuum. The colour charges of the new quarks is causing the string to break and form new hadrons, which are in turn drifting apart, thereby forming a new string. This process continues until the invariant mass of the hadrons is too small to further decay.

The *Cluster Hadronization* [93, 121] model, implemented in HERWIG++ [111], starts by splitting all gluons split into $q\bar{q}$ pairs, leaving only coloured final states. The history of these coloured objects is then traced back through the parton shower and colour connected quarks are paired up to form a colourless cluster. In the next step, the few heavy clusters are split up into lighter clusters by creating more $q\bar{q}$ pairs, thereby forming hadrons.

2.3.4 Underlying Event

Additional interactions, not associated with the hard scattering process of interest are summarized under the term *underlying event* (UE) [93, 122]. The UE includes particles originating from the hadronization of beam-beam remnants (BBR) or multiple parton interactions (MPI), as well as their initial state and final state radiation. In BBR spectator quarks which did not exchange any significant momentum in the collision, are fragmenting and produce hadrons, whereas MPI are additional parton-parton scatterings occurring along with the hard scattering in the same hadron-hadron collision, though with reduced momentum ($p_T \cong 3$ GeV).

The UE models implemented in MC generators require tuning to experimental data, as these effects dominantly produce particles in the low p_T spectrum and therefore cannot be described by pQCD. A set of parameters determined from measurements sensitive to different aspects of the UE is generally referred to as a *tune*.

2.3.5 Detector Simulation

In order to calculate a cross section from the detector level measurement it is necessary to have an estimation of the reconstruction efficiencies and the misidentification rates of particles. Therefore, full events are generated, using the techniques described above and the response of the detector to the traversing particles is simulated using tools like GEANT4 [123].

In the case of CMS (see Sect. 3.2), a detector simulation based on GEANT4 is implemented in the general CMS software framework, including the full detector geometry, magnetic field information and readout performance [124]. An emulation of the CMS trigger system is also included.

To ensure good agreement between the observed spectra in data and simulation, the simulated electronic and hadronic interactions between the particles and the detector material has been validated extensively, using test beam data.

2.4 QCD Analysis Tool: xFitter

XFITTER [125, 126], is a unique open-source QCD fit framework to determine PDFs or fundamental QCD parameters, such as the strong coupling constant α_S or quark masses (m_c, m_b). A schematic overview over the functionalities and the work flow of a QCD fit with XFITTER is presented in Fig. 2.17. The framework is capable of handling data from both fixed target and collider experiments. This includes measurements from ep, $p\bar{p}$ and pp collisions, probing a wide range of Bjorken-x, as is shown in Fig. 2.18.

PDFs of the valence quarks $(xu_v(x), xd_v(x))$, sea-quarks $(x\bar{u}(x), x\bar{d}(x), x\bar{s}(x))$ and gluon $xg(x)$ are parametrized at the starting scale μ_0^2 using a flexible form like the one presented in Eq. 2.49, with different functions used for $P_j(x)$.

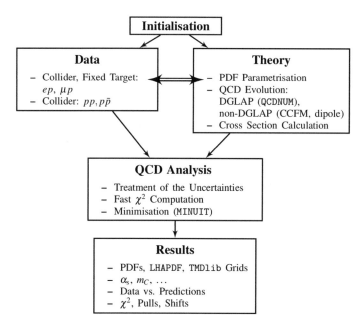

Fig. 2.17 Schematic overview over the xFitter functionalities and work flow [125]

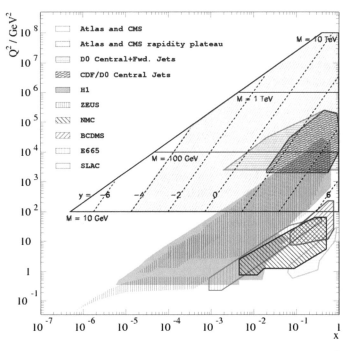

Fig. 2.18 Schematic representation of different regions of phase space in the (x, Q^2) kinematic plane. The capabilities of different colliders and fixed-target experiments to cover these regions are marked with different colours [127]

$$xf(x) = A_j x^{B_j} (1-x)^{C_j} P_j(x) \tag{2.49}$$

$$\text{with:}\quad P_j(x) = 1 + \epsilon_j \sqrt{x} + D_j x + E_j x^2 \tag{2.50}$$

$$\text{or:}\quad P_j(x) = e^{a_3 x} (1 + e^{a_4} x + e^{a_5} x^2) \tag{2.51}$$

The QCD number and momentum sum rules are employed to determine the normalization parameters A_{u_v}, A_{d_v} and A_g. In many analyses, an additional constraint is applied on the strange quark distribution, as they only occur as sea quarks in the proton and many measurements lack the precision to resolve a possible asymmetry:

$$\int_0^1 dx \left[s(x, \mu^2) - \bar{s}(x, \mu^2) \right] = 0 \tag{2.52}$$

The B parameters regulate the behaviour towards small values of x, whereas the C parameter is responsible for the shape of the PDF in the region $x \to 1$. Due to kinematic limitations of datasets used in a QCD analysis, it may be required to restrict the behaviour of parameters, such as $B_{\bar{d}} = B_{\bar{s}}$ or fixing them at particular values, to ensure a convergence of the fit or evaluate parametrization uncertainties.

Among the groups providing global PDFs, HERAPDF [8], MMHT or NNPDF are using polynomials $P_j(x)$ in the style of 2.50, whereas other groups like the CTEQ collaboration prefer polynomials in the form of 2.51 in their parametrizations. The latter cannot assume negative values, by construction. The use of other parametrizations, like the *Big-Log Normal Distributions* or the Chebyshev-polynomials are also supported by XFITTER. Any chosen parametrization requires enough free parameters to assure the fit's flexibility by responding to the used measurements. However, too many free parameters can lead to bias in the sense that the fit can have several local minima or does not converge within a reasonable number of iterations.

The choice of the starting scale is arbitrary, though it is often chosen as $\mu_0^2 = m_c^2$, or slightly below. Once the PDFs have been determined they can be evolved to higher scales $\mu^2 > \mu_0^2$ by different evolution schemes. The default setting of XFITTER is to use the DGLAP equations [42–47], implemented in QCDNUM [128], but the CCFM evolution scheme [52–55], as implemented in uPDFevolve [129] is available, as well.

2.4.1 Heavy Quark Treatment

Different schemes for the treatment of heavy quark masses ($m_h = m_c, m_b$) are implemented in XFITTER. In the *Zero Mass Variable Flavour Number* (ZM-VFN) scheme [130], heavy quarks become available as partons inside the proton when μ^2 reaches the production threshold of $\sim m_h$. Both light and heavy quarks are treated as massless in both the initial and the final state. This approach is valid in the region of $\mu^2 \gg m_h$ and becomes inaccurate for the low-μ^2 regions, due to missing corrections of the order m_h^2/μ^2.

The *Fixed Flavour Number* (FFN) scheme [131–133] uses a fixed number of active flavours in the proton. These can be $n_f = 3$, $n_f = 4$ or $n_f = 5$ depending on the considered heavy quarks, which are treated as an active flavour at a scale far beyond their own production thresholds. The number of flavours is normally set to $n_f = 3$, but it is also possible to set n_f to 4 or even 5, where only the bottom or top quark is treated as heavy. All heavy quarks are produced perturbatively in the final state. This scheme works best in the region of μ^2 close to m_h, as it does not resum terms of the form $ln(\mu^2/m_h^2)$, leading to increasing uncertainties in the fixed order expansion of α_S.

The *general mass variable flavour number* (GM-VFN) aims to combine both schemes by treating the heavy quarks in the FFN scheme at $\mu^2 \approx m_h$ and transitioning to the massless scheme for large scales $\mu^2 \gg m_h$. Different approaches exist to handle the interpolation between the two schemes.

The scheme used in this thesis, the so called Thorne–Roberts [134, 135] GM-VFN requires the descriptions of $n_f = n$ for the FFNS and $n_f = n + 1$ flavour for the GM-VFNS scheme to be equivalent above the transition point between the different schemes. Also, the descriptions must be identical for the transition from $n_f = 1$ to a higher number of available quark flavours $n_f = n + 1$. These conditions are imposed trough the use of coefficient functions C^{FFN}/C^{VFn+1} which differ for particular orders of pQCD.

2.4.2   Methods of Error Estimation in xFitter

The PDFs are fitted by comparing measured cross sections from different processes with theoretical calculations determined at a particular order of pQCD, using
MINUIT [136]. The quality of the fit is evaluated by a χ^2 function, taking into account
the experimental uncertainties of the measured cross sections. For each iteration, the
model parameters are adjusted until a minimum of the χ^2 distribution is reached.

XFITTER offers two options for the treatment of correlated and uncorrelated measurement uncertainties. One is the *covariance matrix* representation, where the χ^2
function can be calculated as follows:

$$\chi^2(m) = \sum_{i,k} (m_i - \mu_i) C^{-1}{}_{ik} (m_k - \mu_k) \tag{2.53}$$

$$\text{with:} \quad C_{ik} = C_{ik}^{\text{stat}} + C_{ik}^{\text{uncor}} + C_{ik}^{\text{sys}}$$

Here, μ_i is the measurement, m_i the theoretical prediction, and C_{ik} the covariance
matrix for measurements in the bins i and k. The latter is a combination of the
statistical, uncorrelated and correlated systematic uncertainties of the experimental
data. However, it is not possible to separate the effects of each individual source of
uncertainty contributing to C_{ik}.

The second option is to use the *nuisance parameter* representation:

$$\chi^2(m, b) = \sum_i \frac{\left[\mu_i - m_i \left(1 - \sum_j \gamma_j^i b_j \right) \right]^2}{\delta_{i,\text{unc}}^2 m_i^2 + \delta_{i,\text{stat}}^2 \mu_i m_i \left(1 - \sum_j \gamma_j^i b_j \right)} + \sum_j b_j^2 , \tag{2.54}$$

where $\delta_{i,\text{stat}}^2$ and $\delta_{i,\text{unc}}^2$ are the relative statistical and uncorrelated systematic uncertainties of the experimental data. $\gamma_j^i b_j$ represents the sensitivity of the measurement
to the correlated systematic source j and b_j denotes the nuisance parameters. In this
definition of χ^2, it is assumed that the systematic uncertainties of the prediction are
proportional to the central values, whereas the statistical uncertainties factor in with
the square root of the expected number of events. The nuisance parameters are determined during the QCD fit, enabling an evaluation of the contributions from different
sources of uncertainties to the result.

Two different methods, the *Hessian method* and the *MC replicas method* are
implemented in XFITTER to propagate the correlated and uncorrelated errors to the
uncertainties of the PDFs. Both methods show good agreement between each other
if the statistical and systematic uncertainties follow a Gaussian distribution [137].
For parton distributions that are not strongly constrained due to a small number of
sensitive measurements or large uncertainties in the fitted datasets, the MC replica
method is expected to give a more realistic picture of the resulting PDF uncertainties.

2.4.2.1 Hessian Uncertainties

The Hessian (Eigenvector) [138] method examines the χ^2 function in the region around the fits minimum χ_0^2. The parameter variations and the resulting PDF uncertainties are evaluated using the Hessian formalism, shown in Eq. 2.55. There, H_{ij} is the Hessian matrix 2.56, which is the matrix of second derivatives of χ^2 at the minimum, with a_i representing the PDF parameters.

$$\Delta\chi^2 = \chi^2 - \chi_0^2 = \sum_{i=1}^{d}\sum_{j=1}^{d} H_{ij}(a_i - a_i^0)(a_j - a_j^0) \qquad (2.55)$$

$$H_{ij} = \frac{1}{2}\frac{\partial^2\chi_{\text{global}}^2}{\partial a_i \partial a_j}\bigg|_0 \qquad (2.56)$$

The parameters determined at χ_0^2 are varied, taking into account a tolerance criterion **T** of $\Delta\chi^2 \leq 1$, corresponding to 68% confidence level. In global PDF fits, using a tolerance criterion this small is normally not possible due to incompatibilities between several datasets used in the fit. This is caused by unknown correlations between the systematic uncertainties of the datasets and the actual measurements not having an ideal statistical behaviour.

The matrix possesses a complete set of orthonormal eigenvectors v_{ik} which can be computed. The obtained eigenvectors correspond to the different, independent sources of uncertainty factoring into the PDFs. These are used to express the displacements around the minimum, as shown in 2.57.

$$a_i - a_i^0 = \sum_{k=1}^{d} v_{ik}s_k z_k \qquad (2.57)$$

There, s_k are normalization scale factors and z_k are the coordinates in the orthonormal eigenvector basis. After the transformation, parameter variations fulfilling the tolerance criterion are enveloped by a hypersphere with radius **T**.

2.4.2.2 MC Replica Method for Uncertainty Estimation

An alternative approach for the treatment of experimental uncertainties in PDF fits is the use of MC-replicas [139, 140]. A sufficient amount of pseudo data is generated ($N_{\text{rep}} > 100$), taking into account the statistical and systematic uncertainties, as well as the correlations for every data point. Each replica is fit individually and the central value of the PDFs is determined as the mean value of all replicas, as shown in Eq. 2.58. The uncertainty is calculated analogue as the Root-Mean-Square(RMS) of the replicas (Eq. 2.59).

$$\langle F \rangle = \frac{1}{N_{\text{rep}}} \sum_{k=1}^{N_{\text{rep}}} F^{(k)} \tag{2.58}$$

$$\sigma^2 \langle F \rangle = \frac{1}{N_{\text{rep}} - 1} \sum_{k=1}^{N_{\text{rep}}} (F^{(k)} - \langle F \rangle) \tag{2.59}$$

2.4.3 Fast Grid Techniques

The determination of higher order pQCD predictions can be very time consuming, making real-time calculations of the necessary theoretical predictions inefficient if they need to be recalculated for each iteration of the fit. To circumvent this issue, different fast grid techniques are available that are approximating the cross section predictions for each point by calculating a sufficient number of *grid points* and using interpolation functions to connect them. This technique has the advantage that the necessary cross section predictions need to be fully calculated only once and any changes in the PDFs or scales (μ_r, μ_f) can be done a posteriori via interpolation.

XFITTER can be interfaced to two fast grid packages, FastNLO [141] and APPLGRID [142]. In the QCD analysis of this thesis, the APPLGRID package is used. Interfacing with programs that provide fixed order calculations for different precesses in pp or p p̄ collisions is possible. The grids are constructed for each bin of the measured distributions, using a two step process. In the first step a scan of the available phase space is done and an optimized number of points in the table is calculated. To provide evenly spaced grid points over the full $x - \mu^2$ range, APPLGRID does not use the direct values for the calculations but a transformation:

$$y(x) = ln\left(\frac{1}{x}\right) + a(1 - x) \qquad \tau(\mu^2) = ln\left(ln\left(\frac{\mu^2}{\lambda_{\text{QCD}}^2}\right)\right) \tag{2.60}$$

With this it is possible to substitute the two PDF terms in the convolution with a suitable approximation, using interpolation functions. The dependence on the scale can be approximated in a similar way, resulting in a three dimensional weight grid. Therefore, any physical observable can be calculated as follows:

$$\sigma = \sum_p \sum_{l=0}^{N_{\text{sub}}} \sum_{i_{y1},i_{y2}} \sum_{i_\tau} W_{i_{y1},i_{y2},i_\tau}^{(p)(l)} \left(\frac{\alpha_S(Q^{2(i_\tau)})}{4\pi}\right)^p F^{(l)}\left(x_1^{(i_{y1})}, x_2^{(i_{y2})}, Q^{2i_\tau}\right) \tag{2.61}$$

Here, N_{sub} is the number of subprocesses contributing to the cross section, $W_{i_{y1},i_{y2},i_\tau}^{(p)(l)}$ are the grid weights and F are the initial state parton combinations.

References

1. Griffiths D (2008) Introduction to elementary particles, 2nd rev. edn. (Physics textbook). Wiley-VCH, Weinheim, 454 p. ISBN: 9783527406012
2. Nagashima Y (2010) Elementary particle physics: Vol. 1: quantum field theory and particles. Wiley-VCH, Weinheim, 942 p. ISBN: 9783527409624
3. Abramowicz H et al (2016) Limits on the effective quark radius from inclusive ep scattering at HERA. Phys Lett B 757:468–472. https://doi.org/10.1016/j.physletb.2016.04.007.. arXiv:1604.01280 [hep-ex]
4. Wikimedia Commons the free media repository. Standard model of elementary particles (2018). https://commons.wikimedia.org/w/index.php?curid=4286964
5. Particle Data Group, Tanabashi M, et al (2018) Review of particle physics. Phys Rev D 98:030001. https://doi.org/10.1103/PhysRevD.98.030001
6. Englert F, Brout R (1964) Broken symmetry and the mass of gauge vector mesons. Phys Rev Lett 13:321-323 [157(1964)]. https://doi.org/10.1103/PhysRevLett.13.321
7. Higgs PW (1964) Broken symmetries and the masses of gauge bosons. Phys Rev Lett 13:508–509 [160(1964)]. https://doi.org/10.1103/PhysRevLett.13.508
8. H1 and ZEUS Collaborations (2015) Combination of measurements of inclusive deep inelastic $e^{\pm}p$ scattering cross sections and QCD analysis of HERA data. Eur Phys J C 75:580. https://doi.org/10.1140/epjc/s10052-015-3710-4. arXiv: 1506.06042 [hep-ex]
9. Thomas AW, Weise W (2001) The structure of the nucleon, 1st edn. Wiley-VCH, Berlin, 389 p. ISBN: 3527402977. https://bibpubdb1.desy.de/record/356952
10. Schörner-Sadenius T (ed) (2015) The Large Hadron Collider: Harvest of run 1. Springer, Cham, 532 p. ISBN: 9783319150000. https://bibpubdb1.desy.de/record/383888
11. Faddeev LD, Popov VN (1967) Feynman diagrams for the Yang-Mills field. Phys Lett B 25:29–30 [325(1967)]. https://doi.org/10.1016/0370-2693(67)90067-6
12. t Hooft G (1973) Dimensional regularization and the renormalization group. Nucl Phys B 61:455–468. https://doi.org/10.1016/0550-3213(73)90376-3
13. Bardeen WA et al (1978) Deep inelastic scattering beyond the leading order in asymptotically free gauge theories. Phys Rev D 18:3998. https://doi.org/10.1103/PhysRevD.18.3998
14. Khachatryan V et al (2017) Measurement and QCD analysis of double-differential inclusive jet cross sections in pp collisions at ps = 8 TeV and cross section ratios to 2.76 and 7 TeV. JHEP 03:156. https://doi.org/10.1007/JHEP03(2017)156. arXiv: 1609.05331 [hep-ex]
15. Bigi IIY et al (1994) The Pole mass of the heavy quark. Perturbation theory and beyond. Phys Rev D 50:2234–2246. https://doi.org/10.1103/PhysRevD.50.2234. arXiv: hep-ph/9402360 [hep-ph]
16. Dowling M, Moch S-O (2014) Differential distributions for top-quark hadro-production with a running mass. Eur Phys J C 74(11):3167. https://doi.org/10.1140/epjc/s10052-014-3167-x. arXiv:1305.6422 [hep-ph]
17. Gray N et al (1990) Three Loop Relation of Quark (Modified) Ms and Pole Masses. Z Phys C 48:673–680. https://doi.org/10.1007/BF01614703
18. Broadhurst DJ, Gray N, Schilcher K (1991) Gauge invariant on-shell Z2 in QED, QCD and the effective field theory of a static quark. Z Phys C 52:111–122. https://doi.org/10.1007/BF01412333
19. Chetyrkin KG, Steinhauser M (1999) Short distance mass of a heavy quark at order_3s. Phys Rev Lett 83:4001–4004. https://doi.org/10.1103/PhysRevLett.83.4001. arXiv: hep--ph/9907509 [hep-ph]
20. Melnikov K, van Ritbergen T (2000) The three loop relation between the MS-bar and the pole quark masses. Phys Lett B 482:99–108. https://doi.org/10.1016/S0370-2693(00)00507-4. arXiv: hep-ph/9912391 [hep-ph]
21. Schmidt B, Steinhauser M (2012) CRunDec: a C++ package for running and decoupling of the strong coupling and quark masses. Comput Phys Commun 183:1845–1848. https://doi.org/10.1016/j.cpc.2012.03.023. arXiv: 1201.6149 [hep-ph]

22. Alekhin S, Moch S (2011) Heavy-quark deep-inelastic scattering with a running mass. Phys Lett B 699:345–353. https://doi.org/10.1016/j.physletb.2011.04.026. arXiv: 1011.5790 [hep-ph]
23. Alekhin S et al (2012) Determination of the charm-quark mass in the MS-bar scheme using charm production data from deep inelastic scattering at HERA. Phys Lett B 718:550–557. https://doi.org/10.1016/j.physletb.2012.11.010. arXiv: 1209.0436 [hep-ph]
24. Alekhin S et al (2013) Precise charm-quark mass from deep-inelastic scattering. Phys Lett B 720:172–176. https://doi.org/10.1016/j.physletb.2013.02.010. arXiv: 1212.2355 [hep-ph]
25. Gizhko A et al (2017) Running of the charm-quark mass from HERA deep-inelastic scattering data. Phys Lett B 775:233–238. https://doi.org/10.1016/j.physletb.2017.11.002. arXiv: 1705.08863 [hep-ph]
26. Abramowicz H et al (2018) Combination and QCD analysis of charm and beauty production cross-section measurements in deep inelastic ep scattering at HERA. Eur Phys J C 78(6):473. https://doi.org/10.1140/epjc/s10052-018-5848-3. arXiv: 1804.01019 [hep-ex]
27. Abdallah J et al (2008) Study of b-quark mass effects in multijet topologies with the DELPHI detector at LEP. Eur Phys J C 55:525–538. https://doi.org/10.1140/epjc/s10052-008-0631-5. arXiv: 0804.3883 [hep-ex]
28. Abramowicz H et al (2014) Measurement of beauty and charm production in deep inelastic scattering at HERA and measurement of the beauty-quark mass. JHEP 09:127. https://doi.org/10.1007/JHEP09(2014)127. arXiv: 1405.6915 [hep-ex]
29. Sirunyan AM et al (2019) Measurement of the tt production cross section, the top quark mass, and the strong coupling constant using dilepton events in pp collisions at ps = 13 TeV. Eur Phys J C 79(5):368. https://doi.org/10.1140/epjc/s10052-019-6863-8. arXiv: 1812.10505 [hep-ex]
30. Aad G et al (2019) Measurement of the top-quark mass in $t\bar{t}$ + 1-jet events collected with the ATLAS detector in pp collisions at ps = 8 TeV. arXiv: 1905.02302 [hep-ex]. http://cds.cern.ch/record/2673558
31. Abe K et al (2017) Search for proton decay via $p \rightarrow e^+\pi^0$ and $p \rightarrow e^+\pi^0$ in 0.31 megaton.years exposure of the Super-Kamiokande water Cherenkov detector. Phys Rev D 95(1):012004. https://doi.org/10.1103/PhysRevD.95.012004. arXiv: 1610.03597 [hep-ex]
32. Placakyte R (2006) First measurement of charged current cross sections with longitudinally polarized positions at HERA. Ph.D thesis, University of München. https://www-h1.desy.de/psfiles/theses/h1th-455.pdf
33. Lichtenberg DB, Rosen SP (eds) (1980) Developments in the quark theory of hadrons, vol. 1, pp 1964–1978
34. Miller G et al (1972) Inelastic electron-proton scattering at large momentum transfers. Phys Rev D 5:528. https://doi.org/10.1103/PhysRevD.5.528
35. Bjorken JD (1969) Asymptotic sum rules at infinite momentum. Phys Rev 179:1547–1553. https://doi.org/10.1103/PhysRev.179.1547
36. Eichten T et al (1973) Measurement of the neutrino - nucleon anti-neutrino - nucleon total cross-sections. Phys Lett B 46:274–280. https://doi.org/10.1016/0370-2693(73)90702-8
37. Brandelik R et al (1979) Evidence for planar events in e+ e annihilation at high-energies. Phys Lett B 86:243–249. https://doi.org/10.1016/0370-2693(79)90830-X
38. Olive KA et al (2014) Review of particle physics. Chin Phys C 38:090001. https://doi.org/10.1088/1674-1137/38/9/090001
39. Bodwin GT (1986) Factorization of the Drell-Yan cross-section in perturbation theory. Phys Rev D 31:2616 [Erratum: Phys Rev D 34:3932 (1986)]. https://doi.org/10.1103/PhysRevD.34.3932,10.1103/PhysRevD.31.2616
40. Collins JC, Soper DE, Sterman GF (1985) Factorization for short distance hadron - hadron scattering. Nucl Phys B 261:104–142. https://doi.org/10.1016/0550-3213(85)90565-6
41. Collins JC, Soper DE, Sterman GF (1988) Soft Gluons and Factorization. Nucl Phys B 308:833–856. https://doi.org/10.1016/0550-3213(88)90130-7
42. Gribov VN, Lipatov LN (1972) Deep inelastic e-p scattering in perturbation theory. Sov J Nucl Phys 15:438

43. Altarelli G, Parisi G (1977) Asymptotic freedom in parton language. Nucl Phys B 126:298. https://doi.org/10.1016/0550-3213(77)90384-4
44. Curci G, Furmanski W, Petronzio R (1980) Evolution of parton densities beyond leading order: the nonsinglet case. Nucl Phys B 175:27. https://doi.org/10.1016/0550-3213(80)90003-6
45. Furmanski W, Petronzio R (1980) Singlet parton densities beyond leading order. Phys Lett B 97:437. https://doi.org/10.1016/0370-2693(80)90636-X
46. Moch S, Vermaseren JAM, Vogt A (2004) The three-loop splitting functions in QCD: the non-singlet case. Nucl Phys B 688:101. https://doi.org/10.1016/j.nuclphysb.2004.03.030. arXiv: hep-ph/0403192 [hep-ph]
47. Vogt A, Moch S, Vermaseren JAM (2004) The three-loop splitting functions in QCD: the singlet case. Nucl Phys B 691:129. https://doi.org/10.1016/j.nuclphysb.2004.04.024. arXiv: hep-ph/0404111 [hep-ph]
48. Aivazis A (2019) Draw Feynman diagram online. https://feynman.aivazis.com
49. Chatrchyan S et al (2017) Measurement of the muon charge asymmetry in inclusive pp ! W+X production at ps = 7 TeV and an improved determination of light parton distribution functions. Phys Rev D 90:032004. https://doi.org/10.1103/PhysRevD.90.032004. arXiv: 1312.6283 [hep-ex]
50. Kuraev EA, Lipatov LN, Fadin VS (1976) Multi-reggeon processes in the Yang-Mills theory. Sov Phys JETP 44:443-450 [Zh Eksp Teor Fiz 71:840 (1976)]
51. Kuraev EA, Lipatov LN, Fadin VS (1977) The pomeranchuk singularity in nonabelian gauge theories. Sov Phys JETP 45:199-204 [Zh Eksp Teor Fiz 72:377 (1977)]
52. Ciafaloni M (1988) Coherence effects in initial jets at small Q2/s. Nucl Phys B 296:49–74. https://doi.org/10.1016/0550-3213(88)90380-X
53. Catani S, Fiorani F, Marchesini G (1990) QCD coherence in initial state radiation. Phys Lett B 234:339–345. https://doi.org/10.1016/0370-2693(90)91938-8
54. Catani S, Fiorani F, Marchesini G (1990) Small x behavior of initial state radiation in perturbative QCD. Nucl Phys B 336:18–85. https://doi.org/10.1016/0550-3213(90)90342-B
55. Marchesini G (1995) QCD coherence in the structure function and associated distributions at small x. Nucl Phys B 445:49-80. https://doi.org/10.1016/0550-3213(95)00149-M. arXiv: hep-ph/9412327 [hep-ph]
56. Aaboud M et al (2018) Measurement of the W-boson mass in pp collisions at ps = 7 TeV with the ATLAS detector. Eur Phys J C 78:110. https://doi.org/10.1140/epjc/s10052-017-5475-4. arXiv: 1701.07240 [hep-ex]
57. Bazarko A.O. et al (1995) Determination of the strange quark content of the nucleon from a next-to-leading order QCD analysis of neutrino charm production. Z Phys C 65:189. https://doi.org/10.1007/BF01571875. arXiv: hepex/9406007 [hep-ex]
58. Goncharov M et al (2001) Precise measurement of dimuon production cross-sections in muon neutrino Fe and muon anti-neutrino Fe deep inelastic scattering at the Tevatron. Phys Rev D 64:112006. https://doi.org/10.1103/PhysRevD.64.112006. arXiv: hep-ex/0102049 [hep-ex]
59. Samoylov O et al (2013) A precision measurement of charm dimuon production in neutrino interactions from the NOMAD experiment. Nucl Phys B 876:339. https://doi.org/10.1016/j.nuclphysb.2013.08.021. arXiv: 1308.4750 [hep-ex]
60. Kayis-Topaksu A et al (2011) Measurement of charm production in neutrino chargedcurrent interactions. New J Phys 13:093002. https://doi.org/10.1088/1367-2630/13/9/093002. arXiv: 1107.0613 [hep-ex]
61. Alekhin S et al (2015) Determination of strange sea quark distributions from fixedtarget and collider data. Phys Rev D 91:094002. https://doi.org/10.1103/PhysRevD.91.094002. arXiv: 1404.6469 [hep-ph]
62. Alekhin S, Blumlein J, Moch S (2012) Parton distribution functions and benchmark cross sections at NNLO. Phys Rev D 86:054009. https://doi.org/10.1103/PhysRevD.86.054009. arXiv: 1202.2281 [hep-ph]
63. Alekhin S, Blümlein J, Moch S (2018) NLO PDFs from the ABMP16 fit. Eur Phys J C 78:477. https://doi.org/10.1140/epjc/s10052-018-5947-1. arXiv: 1803.07537 [hep-ph]

64. Dulat S et al (2016) New parton distribution functions from a global analysis of quantum chromodynamics. Phys Rev D 93:033006. https://doi.org/10.1103/PhysRevD.93.033006. arXiv: 1506.07443 [hep-ph]
65. Harland-Lang LA et al (2015) Parton distributions in the LHC era: MMHT 2014 PDFs. Eur Phys J C 75:204. https://doi.org/10.1140/epjc/s10052-015-3397-6. arXiv: 1412.3989 [hep-ph]
66. Ball RD et al (2017) Parton distributions from high-precision collider data. Eur Phys J C 77:663. https://doi.org/10.1140/epjc/s10052-017-5199-5. arXiv: 1706.00428 [hep-ph]
67. Chatrchyan S et al (2014) Measurement of associated W+charm production in pp collisions at ps = 7 TeV. JHEP 02:013. https://doi.org/10.1007/JHEP02(2014)013. arXiv: 1310.1138 [hep-ex]
68. Catani S, Grazzini M (2007) An NNLO subtraction formalism in hadron collisions and its application to Higgs boson production at the LHC. Phys Rev Lett 98:222002. https://doi.org/10.1103/PhysRevLett.98.222002. arXiv: hep-ph/0703012 [hep-ph]
69. Catani S et al (2009) Vector boson production at hadron colliders: a fully exclusive QCD calculation at NNLO. Phys Rev Lett 103:082001. https://doi.org/10.1103/PhysRevLett.103.082001. arXiv: 0903.2120 [hep-ph]
70. Li Y, Petriello F (2012) Combining QCD and electroweak corrections to dilepton production in FEWZ. Phys Rev D 86:094034. https://doi.org/10.1103/PhysRevD.86.094034. arXiv: 1208.5967 [hep-ph]
71. Placakyte R (2016) Introduction to parton distribution functions. Lecture, CTEQ/MCnet School. https://indico.desy.de/indico/event/13506/contribution/10/material/slides/0.pdf
72. Aad G et al (2012) Determination of the strange quark density of the proton from ATLAS measurements of the $W \to \ell\nu$ and $Z \to \ell\ell$ " cross sections". Phys Rev Lett 109:012001. https://doi.org/10.1103/PhysRevLett.109.012001. arXiv: 1203.4051 [hep-ex]
73. Aaboud M et al (2017) Precision measurement and interpretation of inclusive W+, W^- and $Z/gamma*$ production cross sections with the ATLAS detector. Eur Phys J C 77:367. https://doi.org/10.1140/epjc/s10052-017-4911-9. arXiv: 1612.03016 [hep-ex]
74. QCD analysis of ATLAS $W^{\pm}$ boson production data in association with jets. Technical report, ATL-PHYS-PUB-2019-016. Geneva: CERN (2019). https://cds.cern.ch/record/2670662
75. Alekhin S, Blümlein J, Moch S (2018) Strange sea determination from collider data. Phys Lett B 777:134–140. https://doi.org/10.1016/j.physletb.2017.12.024. arXiv: 1708.01067 [hep-ph]
76. Towell RS et al (2001) Improved measurement of the anti-d/anti-u asymmetry in the nucleon sea. Phys Rev D 64:052002. https://doi.org/10.1103/PhysRevD.64.052002. arXiv: hep-ex/0103030 [hep-ex]
77. Berger EL et al (1989) Weak-boson production at Fermilab Tevatron energies. Phys Rev D 40:83 [Erratum: Phys Rev D 40:3789 (1989)]. https://doi.org/10.1103/PhysRevD.40.83, https://doi.org/10.1103/PhysRevD.40.3789
78. Baur U et al (1993) The Charm content of W + 1 jet events as a probe of the strange quark distribution function. Phys Lett B 318:544–548. https://doi.org/10.1016/0370-2693(93)91553-Y. arXiv: hep-ph/9308370 [hep-ph]
79. Giele WT, Keller S, Laenen E (1996) W plus heavy quark production at the Tevatron. Nucl Phys Proc Suppl C 51:255-260 [255(1996)]. https://doi.org/10.1016/S0920-5632(96)90033-X. arXiv: hep-ph/9606209 [hep-ph]
80. Lai HL et al (2007) The Strange parton distribution of the nucleon: global analysis and applications. JHEP 04:089. https://doi.org/10.1088/1126-6708/2007/04/089. arXiv: hep-ph/0702268 [hep-ph]
81. Alwall J et al (2014) The automated computation of tree-level and next-to-leading order differential cross sections, and their matching to parton shower simulations. JHEP 07:079. https://doi.org/10.1007/JHEP07(2014)079. arXiv: 1405.0301 [hep-ph]
82. Aaltonen T et al (2008) First measurement of the production of aWboson in association with a single charm quark in $p\bar{p}$ collisions at ps = 1.96 TeV. Phys Rev Lett 100:091803. https://doi.org/10.1103/PhysRevLett.100.091803. arXiv: 0711.2901 [hep-ex]

83. Aaltonen T et al (2013) Observation of the production of a W boson in association with a single charm quark. Phys Rev Lett 110:071801. https://doi.org/10.1103/PhysRevLett.110. 071801. arXiv: 1209.1921 [hep-ex]

84. Abazov VM et al (2008) Measurement of the ratio of the $p\bar{p}$! W + c- jet cross section to the inclusive $p\bar{p}$! W+ jets cross section. Phys Lett B 666:23. https://doi.org/10.1016/j.physletb. 2008.06.067. arXiv: 0803.2259 [hep-ex]

85. Aad G et al (2014) Measurement of the production of aWboson in association with a charm quark in pp collisions at ps = 7 TeV with the ATLAS detector. JHEP 05:068. https://doi.org/ 10.1007/JHEP05(2014)068. arXiv: 1402.6263 [hep-ex]

86. Campbell JM, Ellis RK (2010) MCFM for the Tevatron and the LHC. Nucl Phys Proc Suppl 205–206:10. https://doi.org/10.1016/j.nuclphysbps.2010.08.011. arXiv: 1007.3492 [hep-ph]

87. Campbell JM, Keith Ellis R (1999) An update on vector boson pair production at hadron colliders. Phys Rev D 60:113006. https://doi.org/10.1103/PhysRevD.60.113006. arXiv: hep--ph/9905386 [hep-ph]

88. Campbell JM, Keith Ellis R (2015) Top quark processes at NLO in production and decay. J Phys G 42:015005. https://doi.org/10.1088/0954-3899/42/1/015005. arXiv: 1204.1513 [hep-ph]

89. Aaron FD et al (2010) Combined measurement and QCD analysis of the inclusive e+- p scattering cross sections at HERA. JHEP 01:109. https://doi.org/10.1007/JHEP01(2010)109. arXiv: 0911.0884 [hep-ex]

90. Bartosik N (2015) Associated top-quark-pair and b-jet production in the dilepton channel at ps = 8 TeV as test of QCD and background to tt+Higgs production. Universität Hamburg, Diss., 2015. Dr. Hamburg: Universität Hamburg, p 312. https://doi.org/10.3204/DESY-THESIS-2015-035. https://bib-pubdb1.desy.de/record/222384

91. Sjöstrand T, Mrenna S, Skands PZ (2008) A brief introduction to PYTHIA 8.1. Comput Phys Commun 178:852. https://doi.org/10.1016/j.cpc.2008.01.036. arXiv: 0710.3820 [hep-ph]

92. Alwall J et al (2011) MadGraph 5: going beyond. JHEP 06:128. https://doi.org/10.1007/ JHEP06(2011)128. arXiv: 1106.0522 [hep-ph]

93. Buckley A et al (2011) General-purpose event generators for LHC physics. Phys Rep 504:145–233. https://doi.org/10.1016/j.physrep.2011.03.005. arXiv: 1101.2599 [hep-ph]

94. Giele WT, Nigel Glover EW (1992) Higher order corrections to jet crosssections in e+ e annihilation. Phys Rev D 46:1980–2010. https://doi.org/10.1103/PhysRevD.46.1980

95. Giele WT, Nigel Glover EW, Kosower DA (1993) Higher order corrections to jet cross-sections in hadron colliders. Nucl Phys B 403:633–670. https://doi.org/10.1016/0550-3213(93)90365-V. arXiv: hepph/9302225 [hep-ph]

96. Catani S, Seymour MH (1997) A general algorithm for calculating jet crosssections in NLO QCD. Nucl Phys B 485:291–419 [Erratum: Nucl Phys B 510,503 (1998)]. https://doi.org/ 10.1016/S0550-3213(96)00589-5, https://doi.org/10.1016/S0550-3213(98)81022-5. arXiv: hep-ph/9605323 [hep-ph]

97. Catani S et al (2002) The Dipole formalism for next-to-leading order QCD calculations with massive partons. Nucl Phys B 627:189–265. https://doi.org/10.1016/S0550-3213(02)00098-6. arXiv: hep-ph/0201036 [hep-ph]

98. Frixione S, Kunszt Z, Signer A (1996) Three jet cross-sections to next-toleading order. Nucl Phys B 467:399–442. https://doi.org/10.1016/0550-3213(96)00110-1. arXiv: hep--ph/9512328 [hep-ph]

99. Frixione S (1997) A general approach to jet cross-sections in QCD. Nucl Phys B 507:295–314. https://doi.org/10.1016/S0550-3213(97)00574-9. arXiv: hep-ph/9706545 [hep-ph]

100. Kosower DA (1998) Antenna factorization of gauge theory amplitudes. Phys Rev D 57:5410–5416. https://doi.org/10.1103/PhysRevD.57.5410. arXiv: hep-ph/9710213 [hep-ph]

101. Kosower DA (2005) Antenna factorization in strongly ordered limits. Phys Rev D 71:045016. https://doi.org/10.1103/PhysRevD.71.045016. arXiv: hep-ph/0311272 [hep-ph]

102. Daleo A, Gehrmann T, Maitre D (2007) Antenna subtraction with hadronic initial states. JHEP 04:016. https://doi.org/10.1088/1126-6708/2007/04/016. arXiv: hep-ph/0612257 [hep-ph]

103. Phaf L, Weinzierl S (2001) Dipole formalism with heavy fermions. JHEP 04:006. https://doi.
     org/10.1088/1126-6708/2001/04/006. arXiv: hep-ph/0102207 [hep-ph]
104. Nason P (2004) A new method for combining NLO QCD with shower Monte Carlo
     algorithms. JHEP 11:040. https://doi.org/10.1088/1126-6708/2004/11/040. arXiv: hep-
     -ph/0409146 [hep-ph]
105. Frixione S, Nason P, Oleari C (2007) Matching NLO QCD computations with Parton Shower
     simulations: the POWHEG method. JHEP 11:070. https://doi.org/10.1088/1126-6708/2007/
     11/070. arXiv: 0709.2092 [hep-ph]
106. Alioli S et al (2010) A general framework for implementing NLO calculations in
     shower Monte Carlo programs: the POWHEG BOX. JHEP 06:043. https://doi.org/10.1007/
     JHEP06(2010)043. arXiv: 1002.2581 [hep-ph]
107. Campbell JM, Tramontano F (2005) Next-to-leading order corrections to Wt production and
     decay. Nucl Phys B 726:109–130. https://doi.org/10.1016/j.nuclphysb.2005.08.015. arXiv:
     hep-ph/0506289 [hep-ph]
108. Giele WT, Keller S, Laenen E (1996) QCD corrections to W boson plus heavy quark production
     at the Tevatron. Phys Lett B 372:141–149. https://doi.org/10.1016/0370-2693(96)00078-0.
     arXiv: hepph/9511449 [hep-ph]
109. Keller S, Laenen E (1999) Next-to-leading order cross-sections for tagged reactions. Phys Rev
     D 59:114004. https://doi.org/10.1103/PhysRevD.59.114004. arXiv: hep-ph/9812415 [hep-
     ph]
110. Collins JC, Wilczek F, Zee A (1978) Low-energy manifestations of heavy particles: application
     to the neutral current. Phys Rev D 18:242. https://doi.org/10.1103/PhysRevD.18.242
111. Bahr M et al (2008) Herwig++ physics and manual. Eur Phys J C 58:639–707. https://doi.
     org/10.1140/epjc/s10052-008-0798-9. arXiv: 0803.0883 [hep-ph]
112. Frederix R (2018) Parton shower matching at NLO. Lecture, DESY Monte Carlo School.
     https://indico.desy.de/indico/event/19968/contribution/11/material/slides/0.pdf
113. Frederix R et al (2011) Scalar and pseudoscalar Higgs production in association with a top-
     antitop pair. Phys Lett B 701:427–433. https://doi.org/10.1016/j.physletb.2011.06.012. arXiv:
     1104.5613 [hep-ph]
114. Campbell JM et al (2015) Top-pair production and decay at NLO matched with Parton showers.
     JHEP 04:114. https://doi.org/10.1007/JHEP04(2015)114. arXiv: 1412.1828 [hep-ph]
115. Frederix R, Frixione S (2012) Merging meets matching in MC@NLO. JHEP 12:061. https://
     doi.org/10.1007/JHEP12(2012)061. arXiv: 1209.6215 [hep-ph]
116. Lisovyi M, Verbytskyi A, Zenaiev O (2016) Combined analysis of charm-quark
     fragmentation-fraction measurements. Eur Phys J C 76:397. https://doi.org/10.1140/epjc/
     s10052-016-4246-y. arXiv: 1509.01061 [hep-ex]
117. Kniehl BA, Kramer G, Pötter B (2001) Testing the universality of fragmentation func-
     tions. Nucl Phys B 597:337–369. https://doi.org/10.1016/S0550-3213(00)00744-6. arXiv:
     hep-ph/0011155 [hep-ph]
118. Sjöstrand T (1984) Jet fragmentation of nearby partons. Nucl Phys B 248:469–502. https://
     doi.org/10.1016/0550-3213(84)90607-2
119. Andersson B et al (1983) Parton fragmentation and string dynamics. Phys Rep 97:31. https://
     doi.org/10.1016/0370-1573(83)90080-7
120. Bowler MG (1981) e+e Production of heavy quarks in the string model. Z Phys C 11:169.
     https://doi.org/10.1007/BF01574001
121. Brock IC, Schoerner-Sadenius T (eds) (2001) Physics at the terascale. Weinheim: Wiley-VCH,
     476 p. ISBN: 9783527410019
122. Khachatryan V et al (2016) Event generator tunes obtained from underlying event and multi-
     parton scattering measurements. Eur Phys J C 76:155. https://doi.org/10.1140/epjc/s10052-
     016-3988-x. arXiv: 1512.00815 [hep-ex]
123. Agostinelli et al (2003) Geant4 - a simulation toolkit. Nucl Instr Meth A 506:250. ISSN:
     0168-9002. https://doi.org/10.1016/S0168-9002(03)01368-8. https://www.sciencedirect.
     com/science/article/pii/S0168900203013688

124. Bayatian et al (2006) CMS physics: technical design report volume 1: detector performance and software. Technical Design Report CMS. There is an error on cover due to a technical problem for some items. Geneva: CERN. https://cds.cern.ch/record/922757

125. Alekhin S et al (2015) HERAFitter. Eur Phys J C 75:304. https://doi.org/10.1140/epjc/s10052-015-3480-z. arXiv: 1410.4412 [hep-ph]

126. xFitter web site (2018) http://www.xfitter.org/xFitter

127. South DM, Turcato M (2016) Review of searches for rare processes and physics beyond the standard model at HERA. Eur Phys J C 76(6):336. https://doi.org/10.1140/epjc/s10052-016-4152-3. arXiv:1605.03459 [hep-ex]

128. Botje M (2011) QCDNUM: fast QCD evolution and convolution. Comput Phys Commun 182:490. https://doi.org/10.1016/j.cpc.2010.10.020. arXiv: 1005.1481 [hep-ph]

129. Hautmann F, Jung H, Taheri Monfared S (2014) The CCFM uPDF evolution uPDFevolv version 1.0.00. Eur Phys J C 74:3082. https://doi.org/10.1140/epjc/s10052-014-3082-1. arXiv: 1407.5935 [hep-ph]

130. Collins JC, Tung W-K (1986) Calculating heavy quark distributions. Nucl Phys B 278:934. https://doi.org/10.1016/0550-3213(86)90425-6

131. Laenen E et al (1992) On the heavy quark content of the nucleon. Phys Lett B 291:325–328. https://doi.org/10.1016/0370-2693(92)91053-C

132. Laenen E et al (1993) Complete $O(\alpha_S)$ corrections to heavy flavor structure functions in electroproduction. Nucl Phys B 392:162–228. https://doi.org/10.1016/0550-3213(93)90201-Y

133. Riemersma S, Smith J, van Neerven WL (1995) Rates for inclusive deep inelastic electroproduction of charm quarks at HERA. Phys Lett B 347:143–151. https://doi.org/10.1016/0370-2693(95)00036-K. arXiv: hep-ph/9411431 [hep-ph]

134. Thorne RS (2006) Variable-flavor number scheme for NNLO. Phys Rev D 73:054019. https://doi.org/10.1103/PhysRevD.73.054019. arXiv: hepph/0601245 [hep-ph]

135. Martin AD et al Parton distributions for the LHC. Eur Phys J C 63:189. https://doi.org/10.1140/epjc/s10052-009-1072-5. arXiv: 0901.0002 [hep-ph]

136. James F, Roos M (1975) Minuit: a system for function minimization and analysis of the parameter errors and correlations. Comput Phys Commun 10:343–367. https://doi.org/10.1016/0010-4655(75)90039-9

137. Dittmar M et al (2009) Parton distributions. arXiv: 0901.2504 [hep-ph]

138. Pumplin J et al (2001) Uncertainties of predictions from parton distribution functions. 2. The Hessian method. Phys Rev D 65:014013. https://doi.org/10.1103/PhysRevD.65.014013. arXiv: hep-ph/0101032 [hep-ph]

139. Giele WT, Keller S (1998) Implications of hadron collider observables on parton distribution function uncertainties. Phys Rev D 58:094023. https://doi.org/10.1103/PhysRevD.58.094023. arXiv: hep-ph/9803393 [hep-ph]

140. Giele WT, Keller SA, Kosower DA (2001) Parton distribution function uncertainties. arXiv: hep-ph/0104052 [hep-ph]

141. Kluge T, Rabbertz K, Wobisch M (2006) FastNLO: fast pQCD calculations for PDF fits. In: Deep inelastic scattering. Proceedings, 14th international workshop, DIS 2006, Tsukuba, Japan, 20–24 Apr 2006, pp 483–486. https://doi.org/10.1142/9789812706706_0110. arXiv: hep-ph/0609285 [hep-ph]. http://lss.fnal.gov/cgi-bin/find_paper.pl?conf-06-352

142. Carli T et al (2010) A posteriori inclusion of parton density functions in NLO QCD final-state calculations at hadron colliders: the APPLGRID project. Eur Phys J C 66:503. https://doi.org/10.1140/epjc/s10052-010-1255-0. arXiv: 0911.2985 [hep-ph]

Chapter 3
The LHC and the CMS Experiment

High-energy particle accelerators are the most precise microscopes in the world, providing a unique opportunity to study the fundamental building blocks of matter. In an accelerator experiment, particles are accelerated and brought to collision, with their products being recorded in complex detector systems. The energy of the collisions determines the maximum resolution of an accelerator, while the luminosity defines the observation rates of the processes of interest. Such experiments are large scale facilities, which can only be operated by big international collaborations.

The Large Hadron Collider (LHC) at CERN is the machine at the very frontier of energies to date and allows for the most detailed picture of the fundamental structure of visible matter. At the LHC, the collisions of protons or heavy ions[1] are recorded and analyzed by 4 complex multi-purpose experiments: ATLAS [2], CMS [3], ALICE [4] and LHCb [5]. ATLAS and CMS are designed to cover a broad spectrum of physics topics, whereas LHCb is specialized in the measurement of CP violation and rare decays of B hadrons. The focus of ALICE is the study of Quark Gluon Plasma (QGP) properties via heavy ion collisions. The work in this thesis considers measurements obtained with the CMS experiment. In this chapter, the LHC and the CMS detector are described.

3.1 The Large Hadron Collider

The Large Hadron Collider (LHC) [6] is a circular collider operating at highest center-of-mass energies, to date. At a circumference of 26.7 km, it is installed at CERN near Geneva in an underground tunnel, originally built for the LEP [7] e^+e^--collider. The LHC machine is designed to accelerate and collide pp, pPb and PbPb beams up to

---

[1]In this thesis, only proton collisions are investigated.

© The Editor(s) (if applicable) and The Author(s), under exclusive license   47
to Springer Nature Switzerland AG 2020
S. K. Pflitsch, *Associated Production of W + Charm in 13 TeV
Proton-Proton Collisions Measured with CMS and Determination of the Strange Quark
Content of the Proton*, Springer Theses,
https://doi.org/10.1007/978-3-030-52762-4_3

CERN's accelerator complex

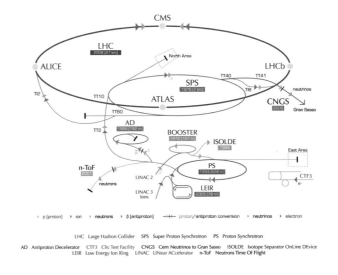

Fig. 3.1 Illustration of the CERN accelerator complex showing the acceleration steps and location of different experiments [8]

a center-of-mass energy of up to $\sqrt{s} = 14\,\text{TeV}$. Starting from a linear accelerator (LINAC-2 for protons, LINAC-3 for heavy ions), the particles pass through different stages of acceleration before they are transferred to the LHC at a beam energy of 450 GeV. An illustration of the CERN accelerator complex and the positions of the detectors belonging to the different experiments hosted at the LHC is presented in Fig. 3.1.

After injection into the LHC, the particles are further accelerated using superconducting radio frequency (RF) cavity systems, which oscillate at a frequency of 400 MHz. The cavities consist of niobium sputtered on copper and a single one can reach a maximum accelerating voltage of 2 MV, corresponding to a total of 16 MV for a full beam. Accelerating particles through oscillating fields causes the protons or heavy ions to form discrete packets, so called *bunches*. At peak performance, a total of 2808 bunches, each carrying $1.15 \cdot 10^{11}$ protons, circulate inside the LHC.

The bunches are kept on a circular trajectory by 1232 superconducting dipole magnets, consisting of NbTi. These are operated at temperatures below 2 K, using superfluid helium as a cooling liquid, and generate fields above 8 T for beam energies of 7 TeV. The beam dispersion caused by the protons Coulomb repulsion is countered by using quadrupole and other higher order magnet systems to focus the beam. For this purpose, 392 superconducting NbTi quadrupoles are installed at the LHC to squeeze the beam either in the vertical or horizontal direction.

3.1.1 LHC Luminosity

The beams of the LHC are intersecting at four interaction points, with bunch crossings occurring every 25ns in the standard Run-II setup. The average number of particle collisions per bunch crossing, and therefore the instantaneous luminosity delivered, depends on several machine parameters and can be written as follows:

$$\mathcal{L}_{\text{inst}} = \frac{N_b^2 \, n_b \, f \, \gamma}{4\pi \, \epsilon_n \, \beta^*} F \qquad (3.1)$$

Here, N_b corresponds to the number of particles per bunch, n_b to the number of bunches, f is the rotation frequency, γ the relativistic gamma-factor, β^* the beta-function at the collision point and ϵ_n is the transverse beam emittance. F is the so called geometric luminosity reduction factor, which is correlated with the crossing angle of the beams at the interaction point.

The CMS and ATLAS experiments work with a high number of collisions per bunch crossing corresponding to an instantaneous luminosity of $L = 10^{34}\,\text{cm}^{-2}\text{s}^{-1}$ at their interaction points, to increase the number of recorded processes with low cross sections. $\mathcal{L}_{\text{inst}}$ is determined by measuring observables that increase proportional to the number of pp collisions per bunch-crossing, such as the number of clusters in the tracking detector of an experiment. Calibrations for the different rates of observables used for the luminosity calculation are obtained via a Van-der-Meer scan [9]. In these scans, the crossing angle and overlap of the beams are shifted in many small steps and the resulting change in the average number of pp collisions is reflected in the rates of the observables. From the calibration data obtained in the Van-der-Meer scan, it is possible to determine the instantaneous luminosity for the regular running conditions and calculate the integrated luminosity. In 2016, the LHC delivered a total of $40\,\text{fb}^{-1}$ during proton-proton operation [10].

3.2 The Compact Muon Solenoid Experiment

The detector of the Compact Muon Solenoid (CMS) experiment is located in an underground cavern at the LHC Point 5 close to Cessy, France. With its proportions of 21.6 m in length, a diameter of 14.6 m and a total weight of 12500 t, it is relatively compact when compared to the detector of the ATLAS experiment. It is designed as a multi-purpose detector, capable of performing ultimate precision tests of Standard Model parameters and searches for physics beyond the Standard Model [11].

The CMS detector is built in a cylindrical shape around the LHC beam pipe, with the different detectors arranged in cylindrical layers in the central region and two endcaps in the forward and backward region. One of the key elements of the experiment is the superconducting solenoid magnet, providing a homogeneous field of 3.8 T. Outside of the solenoid, the magnetic field is returned through an iron yoke,

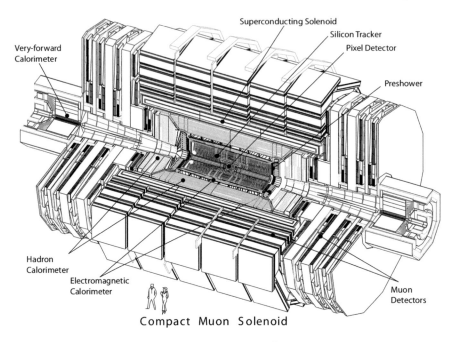

Superconducting Solenoid
Silicon Tracker
Pixel Detector
Very-forward
Calorimeter
Preshower
Hadron
Calorimeter
Electromagnetic
Calorimeter
Muon
Detectors

Compact Muon Solenoid

Fig. 3.2 Illustration of the CMS detector with all of the different components [1]

consisting of 5 rings with a width of 2.536 m and a thickness of 19.5 cm, and 2 endcaps, comprised of three disks each. Several components of the muon system are embedded in the yoke and it is used as an additional hadron absorber. The different components of the detector are presented in Fig. 3.2. The ones that are most relevant for the physics analysis addressed in this thesis are described in the next sections.

3.2.1   The CMS Coordinate System

The centre of the CMS coordinate system is located at the nominal collision point inside of the detector. Its xy-plane is oriented perpendicular to the beam, with the x axis pointing towards the centre of the LHC and the y-axis pointing upward. When using a cylindrical representation of the coordinates, the azimuthal angle ϕ and the radial coordinate are defined in the xy-plane, with ϕ measured in reference to the x-axis. The z-axis is oriented in the direction of the beam and the polar angle θ is measured in reference to this axis.

The transverse momentum of a particle is calculated from the x and y components of the recorded data and is defined as $p_{\mathrm{T}} = \sqrt{p_x^2 + p_y^2}$. Another commonly used observable is pseudorapidity, which is invariant under Lorentz transformation for particles with large momenta ($|p| \gg m$) and is defined as $\eta = -ln[tan(\theta/2)]$.

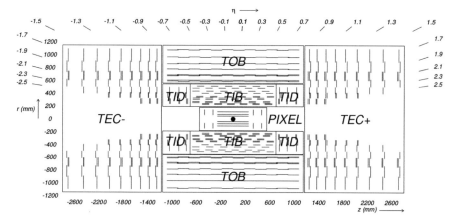

Fig. 3.3 Schematic cross section of the CMS tracker. The lines represent the detector modules. Double lines indicate back-to-back modules delivering stereo hits [3]

With this, the four-vector of a particle can be expressed through a combination of (E, p_T, η, ϕ), as well as (E, p_x, p_y, p_z).

3.2.2 The Tracker

The tracking system is designed to measure the trajectories of charged particles, produced in hadron collisions, and to reconstruct primary vertices. Secondary vertices from the decays of long lived heavy particles can be reconstructed as well. The tracker is the main component of the CMS detector, used in the reconstruction and identification of electrons and muons, as well as several decay topologies of tau leptons involving charged hadrons. The Tracker has a barrel region and two endcaps, each consisting of several layers of silicon pixel and silicon strip detectors, covering part of the region of $|\eta| < 2.5$. Due to the high particle flux in the area closest to the beam, it is the detector most affected by radiation damage. Operating under the LHC conditions, producing on average 1000 particles per bunch-crossing, requires a fine granularity and fast response time to ensure a good separation of the tracks.

An illustration of the cross section of the tracking detector is presented in Fig. 3.3. The pixel detector contains 1440 modules, arranged in 3 layers around the beam pipe, as well as 2 layers of modules in both endcap disks. A total of 66 million pixels are installed within the modules, each $100 \times 150\,\mu m$ in size and with a thickness of $285\,\mu m$. This setup reaches a spatial resolution of $10\,\mu m$ in the transversal coordinate and $20\text{-}40\,\mu m$ in the longitudinal coordinate [12].

The strip detector surrounds the pixel detector and consists of 15148 modules with a total of 9.3 million strips. The modules are arranged in 10 barrel-like layers and 12 disk-like layers in the endcaps, separating the tracker into an inner and an outer region. The strip thickness is $320\,\mu$m for all layers in the inner region and several of the endcap layers in the outer regions. The remaining layers use strips with a thickness of $500\,\mu$m. The spatial resolution for the inner layers is approximately 13–$38\,\mu$m, whereas it reaches approximately 18–$47\,\mu$m in the outer region.

3.2.3   The Electromagnetic Calorimeter

The electromagnetic calorimeter (ECAL) [13] measures the energy of electrons and photons. In combination with the information provided by the tracker, it plays an important role in particle identification.

The detector is positioned inside the solenoid and surrounds the tracker. It consists of 76000 lead tungstate (PbWO$_4$) crystals, arranged in a barrel-like geometry with a pseudorapidity coverage of $|\eta| < 1.5$. The two endcap calorimeters on each side, provide a total coverage of up to $|\eta| = 3$. The ECAL belongs to the class of homogeneous calorimeters, meaning that the crystals function as both absorber and scintillator at the same time. The face cross section of each crystal is 22×22 mm^2 in the barrel region and 28.6×28.6 mm^2 in the endcap disks. Their length is 230 mm, corresponding to 25.8 radiation lengths and the Molière radius of the material is 2.2 cm. In the barrel region avalanche photo-diodes are glued to the back of each crystal to collect the scintillation light produced by traversing particles, whereas vacuum photo triodes are used in the endcaps.

A preshower detector is installed in front of each of the ECAL endcaps in the region of $1.65 < |\eta| < 2.61$. It belongs to the class of sampling calorimeters and consists of alternating layers of lead radiators and silicon strip sensors. Its main purpose is the identification of photons coming from $\pi^0 \rightarrow \gamma\gamma$ decays against prompt photons. An improvement in the position determination of electrons and photons is achieved as well.

3.2.4   The Hadronic Calorimeter

The hadronic calorimeter (HCAL) [14] is essential for the measurement of hadronic jets and the calculation of missing transverse momentum, which indicates the presence of neutrinos. It is designed as a sampling calorimeter and consists of four subdetectors, the barrel (HB), the outer barrel (HO), the endcap disks (HE), and the forward calorimeters (HF).

The barrel covers the region of $|\eta| < 1.4$ and is located inside the solenoid together with the two endcaps, which cover the region of $1.3 < |\eta| < 3.0$. The HB consists of 36 wedges with alternating layers of brass absorber plates (56.5 mm thick) and plastic scintillator plates (3.7 mm thick). In the most central region, the absorber thickness of the material corresponds to 5.82 interaction lengths (λ_I) and increases with the polar angle θ, reaching $10.6\,\lambda_I$ at $|\eta| = 1.3$. The absorber plates of the endcaps are larger (97 mm thickness), with 9 mm gaps for the scintillator plates. This results in an absorber thickness of approximately $10\,\lambda_I$ throughout the HE. A total of 70000 scintillator tiles are integrated in the HCAL and connected to multi-channel hybrid photo diodes through wavelength-shifting fibres for readout.

The outer HO is used to identify late starting showers and provides an improved energy resolution for high p_T jets in the central region. Due to spatial limitations it is placed outside the magnet, covering the region $|\eta| < 1.26$. It uses the solenoid itself as absorber and 10 mm thick scintillator plates are placed before the muon system. Due to the low absorber depth in the very central region, two layers of HO scintillator are placed on either side of a 19.5 cm thick piece of iron.

Additional forward calorimeters are located at $3.0 < |\eta| < 5.0$ to ensure the hermeticity of the energy measurement. They consist of 5 mm thick steel plates with embedded quarz-fibres. This design was chosen due to the requirements on radiation hardness for the active medium in this region. The average energy deposit in this component is 760 GeV in comparison to the 100 GeV which are expected for rest of the detector. The HF is most sensitive to the electromagnetic component of hadronic showers, as it detects Cherenkov radiation emitted by traversing charged particles above the Cherenkov threshold ($E \geq 190$ keV for electrons).

3.2.5 The Muon System

The muon system [15] is the main component of the CMS detector and has been designed for the identification, momentum measurement, and triggering of muons. As muons are less affected than electrons by radiative losses in the tracker material, using these particles in event reconstruction provides a better mass resolution. To account for the shape of the solenoid magnet and different background conditions, three types of gaseous particle detectors are installed in the CMS muon system, as illustrated in Fig. 3.4.

The CMS barrel muon detector covers the region of $|\eta| < 1.2$ where the muon flux is low and the magnetic field is relatively uniform. There, drift tubes (DT) are installed in 4 concentric cylinders around the beam pipe, with two embedded in the iron yoke, one placed before it, and one on the outside. The three inner cylinders contain 60 drift chambers and each chamber consists of 3 super layers (SL). In the first two layers the drift wires are aligned parallel to the beam line to measure the r-ϕ coordinate, whereas the wires in the third layer are aligned orthogonal to the beam line for the measurement of the z-coordinate. The fourth cylinder contains 70 drift chambers with only two super layers, providing r-ϕ information only. Each super layer in turn consists of four layers of rectangular drift cells. The drift cells are filled with a gas mixture consisting of 85% Ar and 15% CO_2 and have a tube cross section of 13×42 mm^2. A traversing muon ionizes the gas in the cell and the freed electrons drift to the cells anode wires where they are registered as a signal. About 172,000 sensitive wires are integrated in the drift cells, each consisting of gold plated stainless steel with a diameter of 50 μm, and are operated at $+3600$ V. The electric

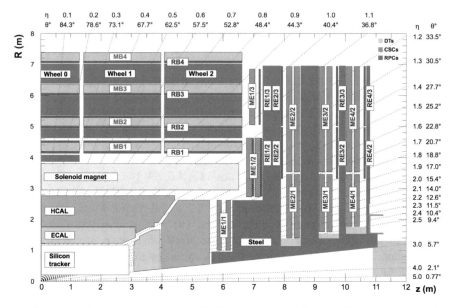

Fig. 3.4 Illustration of one quarter of the CMS muon system. Drift tubes (DT) are utilized in the central barrel region and cathode strip chambers (CSC) are used in the endcap regions. Resistive Plate Chambers are placed in between for improved momentum resolution and trigger capabilities [15]

field inside the DTs is relatively uniform, thereby ensuring an approximately constant drift velocity of the electrons so the hit position of the muon can be determined by a time of flight measurement. This configuration provides a spatial resolution between 250–300 µm for layers measuring the ϕ-coordinate and 600 µm for layers measuring the θ-coordinate.

The two endcaps of the muon detector cover the region of $0.9 < |\eta| < 2.4$, where high muon rates and a non-uniform magnetic field have to be taken into account. For these reasons, cathode strip chambers (CSC) were chosen for the design. CSCs are multiwire proportional chambers consisting of alternating layers of anode wire arrays and cathode panels inside a gas volume. The gas mixture chosen for the CMS detector consists of 50% CO_2, 40 Ar, and 10% CF_4. The wires are aligned azimuthally, defining a track's radial coordinate. Strips are milled on the cathode panels, approximately perpendicular to the wires. The muon coordinate along the wires is measured by interpolating charges induced on strips. Four CSC stations are installed in each endcap, with the chambers positioned perpendicular to the beam line, embedded in between the flux return plates. The spatial resolution of the CSCs is 75 µm for the station closest to the interaction point and increases to 150 µm for the one furthest away.

Both, the DT and the CSC are capable of triggering on the muon p_T with good efficiency. Due to the high background rates and an inefficiency in measuring the correct beam-crossing time when the LHC reaches full luminosity, a complementary, dedicated trigger system is installed in the detector. Resistive plate chambers (RPC) are embedded in the barrel iron yoke as well as the endcaps, as is shown in Fig. 3.4. The chambers are filled with a gas mixture consisting of 95.2% $C_2H_2F_4$, 4.5% i-C_4H_{10}, 0.3% SF_6. Water vapour is added to the mixture until it reaches a relative humidity of 40–50%. The RPCs are double-gap chambers, operated in avalanche mode to ensure good operation at high rates. These consist of high resistance plates, separated by a 2 mm gas-filled volume with pick-up readout strips in between. RPCs provide spatial resolution, a timing resolution much faster than the 25 ns in between two bunch crossings, and a good estimation of the transverse momentum of the traversing particle.

3.2.6 The Trigger System

At full capacity the bunch crossing rate of the LHC reaches 40 MHz [1, 16], resulting in about 10^9 pp collisions per second at the CMS interaction point. Since it is impossible to read out and store all of the collision data, triggers are employed to reduce the rate to a manageable level. These also act as filters for interesting event signatures to be used in later physics analyses. The decision to read out the full event data is reached via a two level system, with the first level (L1) consisting of custom hardware triggers and the higher level (HLT) is a processor farm.

For the L1, information from the ECAL, HCAL and the muon system are processed and factor into the decision of the global trigger to accept or reject an event. Events are selected via candidate objects above pre-set thresholds, so called "trigger primitives" that are working with reduced granularity and resolution. These can be large energy deposits in the calorimeters, indicating the presence of an electron or a photon, or ionization deposits in the muon system, indicating the presence of a muon. The L1 decision is reached within 4 ns of a bunch crossing, where under 1 ns is required for the trigger calculations and the remainder is needed for the data transit. During the transfer, the high-resolution data from the detector is kept in buffers to have as little dead time as possible during operation.

The HLT hardware is a processor farm with ~13000 CPUs performing event selection, based on offline-quality reconstruction algorithms. Several processing steps are run on the data in a predefined order, a so called *HLT-path*, reconstructing physics objects and making selections based on these objects. The complexity of the algorithms and reconstruction refinement increase with each step, making it sensible to first handle the data from the calorimeters and muon system before performing the more sophisticated track reconstruction.

Events that have been accepted by the HLT are archived by a software process called *storage manager*. The recorded data is stored on disk and transferred to the CMS Tier-0 computing centre for further processing and permanent storage.

References

1. Bayatian et al (2006) CMS physics: technical design report volume 1: detector performance and software. Technical design report CMS. There is an error on cover due to a technical problem for some items. Geneva: CERN. https://cds.cern.ch/record/922757
2. The ATLAS Collaboration (2008) The ATLAS experiment at the CERN large hadron collider. J Instrum 3(08):S08003. http://stacks.iop.org/1748-0221/3/i=08/a=S08003
3. The CMS Collaboration (2008) The CMS experiment at the CERN LHC. J Instrum 3(08):S08004. http://stacks.iop.org/1748-0221/3/i=08/a=S08004
4. The ALICE Collaboration (2008) The ALICE experiment at the CERN LHC. J Instrum 3(08):S08002. http://stacks.iop.org/1748-0221/3/i=08/a=S08002
5. The LHCb Collaboration (2008) The LHCb detector at the LHC. J Instrum 3(08):S08005. http://stacks.iop.org/1748-0221/3/i=08/a=S08005
6. Evans L, Bryant P (2008) LHC machine. J Instrum 3(08):S08001. http://stacks.iop.org/1748-0221/3/i=08/a=S08001
7. Camilleri et al (1976) Physics with very high-energy e+e colliding beams. CERN Yellow Reports: Monographs. Geneva: CERN. https://cds.cern.ch/record/185961
8. Lefèvre C (2008) The CERN accelerator complex. Complexe des accélèrateurs du CERN. https://cds.cern.ch/record/1260465
9. van der Meer S (1968) Calibration of the effective beam height in the ISR. Technical report CERN-ISR-PO-68-31. ISR-PO-68-31. Geneva: CERN. https://cds.cern.ch/record/296752
10. The CMS Collaboration (2017) CMS luminosity measurements for the 2016 data taking period. CMS physics analysis summary CMS-PAS-LUM-17-001. http://cds.cern.ch/record/2257069
11. Bayatian et al (2007) CMS physics: technical design report volume 2: physics performance. J Phys G 34.CERN-LHCC-2006-021. CMS-TDR-8-2 (2007) revised version submitted on 22 Sep 2006 17:44:47, 995–1579. 669 p. https://cds.cern.ch/record/942733
12. The CMS Collaboration (2014) Description and performance of track and primary vertex reconstruction with the CMS tracker. J Instrumen 9(10):P10009. http://stacks.iop.org/1748-0221/9/i=10/a=P10009
13. The CMS electromagnetic calorimeter project: technical design report (1997) Technical design report CMS. Geneva: CERN. http://cds.cern.ch/record/349375
14. The CMS hadron calorimeter project: technical design report (1997) Technical design report CMS. Geneva: CERN. https://cds.cern.ch/record/357153
15. Sirunyan AM et al (2018) Performance of the CMS muon detector and muon reconstruction with proton-proton collisions at ps = 13 TeV. JINST 13:P06015. https://doi.org/10.1088/1748-0221/13/06/P06015. arXiv:1804.04528 [physics.ins-det]
16. Khachatryan V et al (2017) The CMS trigger system. JINST 12:P01020. https://doi.org/10.1088/1748-0221/12/01/P01020.. arXiv:1609.02366 [physics.ins-det]

Chapter 4
Event Reconstruction

All physics objects, including tracks, jets or muons that are to be used in physics analyses are reconstructed from the signals recorded in the different components of the detector. An illustration of the signals recorded by the CMS detector, originating from different particle types is presented in Fig. 4.1. In this chapter, the algorithms used in the CMS experiment for the reconstruction of physics objects are discussed, with a focus on the ones needed in the W + c measurement.

4.1 Track Reconstruction

Tracks and vertices are reconstructed using the *combinatorial Kalman fitter* [3–5], as implemented in the software package *CMS combinatorial track finder* [6, 7]. Tracks and vertices are fitted from the reconstructed hits in the different layers of the tracker to measure the trajectories, charge sign and momentum of charged particles. Under the LHC running conditions, it is necessary to balance the need for a high track reconstruction efficiency, a low fake-rate and the processing speed of the algorithms, as it is included in HLT decisions. With the CMS tracking algorithms, it is possible to reconstruct tracks going down to 0.1 GeV in p_T inside the full η-range of the tracker.

An important parameter used in the reconstruction of tracks is the so called *beam spot* [8], which is a 3D profile of the CMS luminous region where the beams of the LHC collide. Its parameters (position, width) are determined as the average over many events and knowledge of its position provides crucial information for the event reconstruction. Especially the beam spot position plays a significant role in the reconstruction of tracks and the determination of their impact parameter, which is defined as the distance of closest approach (DCA) in the xy-plane between the track and its origin, in this case the primary vertex.

S. K. Pflitsch, *Associated Production of W + Charm in 13 TeV
Proton-Proton Collisions Measured with CMS and Determination of the Strange Quark
Content of the Proton*, Springer Theses,
https://doi.org/10.1007/978-3-030-52762-4_4

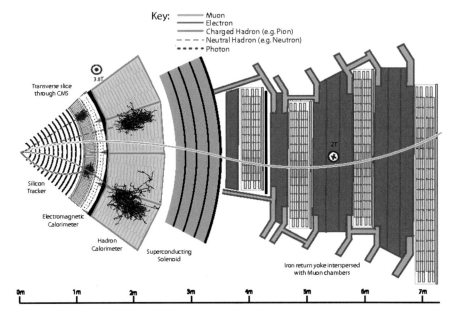

Fig. 4.1 Illustration of a transverse slice of the CMS detector. Interactions of different types of particles with the subdetectors are shown as lines and clusters of different colours. The muon and the charged pion have positive charges and the electron is negatively charged [2]

The tracks of an event are reconstructed via an iterative fitting procedure, described in the following. The first iteration, referred to as *iteration 0* reconstructs prompt tracks which originated close to the pp-collision with $p_T^{track} > 0.8$ GeV and 3 pixel hits. At the next iteration, *iteration 1*, prompt tracks with only two pixel hits are reconstructed, though the minimum p_T of the candidates remains unchanged. For *Iteration 2*, the p_T requirement is released and *iterations 3–5* aim to find tracks originating outside the beamspot.

Each iteration follows four steps. Step 1: The track seed is generated to have an initial estimate of the trajectory and the associated uncertainties. It consists of 2–3 hits in the tracker layers. In the first iteration, only seeds consisting of three hits in the Pixel detector, consistent with a prompt track are considered, as they provide well measured starting trajectories. In further iterations, seeds with only two pixel or strip hits are considered, the p_T^{track} thresholds of the track candidates are lowered, and the track origin is allowed to be further away from the beam spot.

Step 2: the trajectories are extrapolated along the expected flight path, searching for more hits that can be assigned to the track candidate. All potential hits close to the expected trajectory are tested using a χ^2 test, taking into account the uncertainties of the current track candidate, as well as uncertainties of the tracker hit. The initial estimate of the track parameters from the hits of the track seed is performed and refined for each hit added to the track candidate.

Step three: All hits associated with a track candidate are refitted in order to reduce bias from the selection criteria of the previous steps. It provides a best-possible estimate on all parameters of the trajectory and possible spurious hits from nearby tracks or electronic noise are removed.

In the fourth step, the track candidates are assigned a quality label, based on the number of different tracker layers contributing to the fit, its χ^2/N_{dof} value and compatibility with originating from a primary vertex. Tracks assigned to the *loose* category fulfill the minimum quality requirements of a track and can be used in analyses with a focus on selection efficiency, where the fake rate is only a minor concern. Tracks assigned to the *tight* or *high purity* category have to fulfill increasingly stricter selection criteria regarding the minimum number of layers with hits and the maximum number of lost layers where no hit could be reconstructed, in order to reduce the fake rate. The selection criteria of the quality labels differ for each iteration of the tracking algorithm to account for the different seed conditions and track finding requirements, and are documented in detail in Ref. [7].

To improve the track reconstruction performance, the alignment [9] of all pixel and strip modules of the tracker is determined with a precision of a few μm. Possible deviations from the presumed module using the recorded cosmics data or dedicated runs in which the magnetic field is off, making use of the particles straight trajectory. The tracks are fitted, using two different, though complementary, approaches to determine the alignment parameters. The first option is a simultaneous fit of all track and alignment parameters, the so called *Millpede II* approach [9, 10], which is taking all parameter correlations into account. The second option is the use of a local fit, the so called *HipPy* approach [10], where the correlations are handled iteratively.

4.2 Primary Vertex Reconstruction

The reconstructed tracks are used to determine the location and associated uncertainties of all pp-interactions in an event. The reconstructed vertex with the largest value of p_{T}^2 for the complete physics-object is taken to be the primary pp interaction vertex and additional pp-collisions from the same (in-time) or adjacent (out-of-time) bunch crossings are generally referred to as *pileup*. The reconstruction of vertices is done in a three-step process, consisting of tracks selection, track clustering and fitting of the vertex position.

A track needs to fulfill the following selection criteria to be considered originating from a primary vertex. The track's transverse impact parameter significance $(\text{d}_{xy}/\sigma\text{d}_{xy})$ relative to the beamspot is required to be below 5. Additionally the track must possess a minimum number of pixel and strip hits associated with it (pixel hits ≥ 2, pixel hits + strip hits ≥ 5) and the normalized χ^2 of the track's trajectory fit must be below $\chi^2/N_{\text{dof}} < 20$. No requirement on the minimum $p_{\text{T}}^{\text{track}}$ is imposed in the selection. Figure 4.2 is an illustration of tracks associated with a secondary vertex from the decay of a long lived B-meson originating from a primary

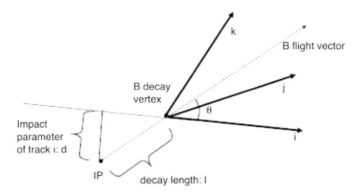

Fig. 4.2 Illustration of the impact parameter of a track produced in the decay of a long lived particle [11]

vertex. The impact parameter of the track is large and would therefore not be selected for the primary vertex fit.

The selected tracks are then clustered according to their z-coordinate and point of closest approach to the centre of the beam spot, for which a deterministic annealing (DA) [12] algorithm is used. Along with the assignment of tracks to a primary vertex, the DA algorithm also determines the number vertices in an event. The algorithm must separate vertices in close proximity, which is important in a high pileup environment, but also consider the possibility that tracks originating at the same vertex are separated.

The candidate vertices are fitted using an adaptive vertex fitter [13] to determine the best estimate on all of the vertex parameters. This includes the vertex position in x, y and z, the covariance matrix and indicators for the success of the fit. It is based on the Kalman fitter, but includes weights for each track, indicating the probability of the track being compatible with the vertex position. The procedure is iterative, starting from an initial vertex position and track weights, re-calculating both until convergence is reached.

4.3 Particle Flow

The particle-flow algorithm [2] aims to provide a global event description. This includes the identification of all particles produced in a collision and discriminates against particles produced in pileup collisions. This is achieved by correlating the signals recorded in all of the subdetectors of CMS via a *link algorithm*. The probability to combine elements originating from a single particle depends on the granularity of the detector and the pileup conditions of the run, whereas the probability to link all elements of a given particle depends on the material encountered on the particles trajectory.

Tracks are extrapolated from their last hits in the tracker to the first layers of the calorimeters or the muon detector to identify electron, muon and charged hadron candidates. A link is established if the extrapolated track position is within an area around the shower maximum, taking into account the calorimeters cell size, possible gaps between the cells and uncertainties in the determination of the shower maximum. In cases where tracks can be combined with multiple clusters from a calorimeter or vice versa, the combination with the smallest distance in the $\eta\phi$-plane is chosen. Other combinations, such as the link between ECAL and HCAL clusters for jets or the combination of two tracks to form a secondary vertex are investigated as well.

4.3.1 Muon Identification

The muon candidate objects are selected from the tracks which are reconstructed by fitting either the hits in the tracker or those in the muon system. These are sorted into 3 categories.

Standalone muons are constructed from information provided by the muon system only. Muons that are assigned to this category, generally have worse momentum resolution than the other categories and the contamination with cosmic muons is higher.

Tracker muons are based on a tracks that were reconstructed in the inner tracker and extrapolated to the muon detector, where it is matched to hits in either the DT or the CSC. Those tracks with $p_T^\mu > 0.5\,\text{GeV}$ and $p_T^\mu > 2.5\,\text{GeV}$ are, respectively, considered as muon candidates. Muons that are only identified as *tracker muons* usually have a low transverse momentum and only hits in the first layer of the muon system. This leads to an increased rate of high-p_T charged hadrons to be misidentified as muons, as these particles can reach the innermost layer of the muon system.

Global muons are constructed by matching a standalone muon track with a track from the inner parts of the detector. If the track of a global muon coincides with the track of a tracker muon, it is refitted, taking into account the hits reconstructed in the tracker as well as the ones from the muon system.

The categorization is then refined, depending on the precision requirements of the muon identification, following the physics task of the analysis. The *loose ID* only requires that the candidate is either a tracker or a global muon. The *soft ID* is optimized for low-p_T muons from B-hadron decays, while the *high momentum ID* is designed for muons with a transverse momentum above $p_T^\mu > 200\,\text{GeV}$.

The *tight ID*, which is used in the W + c analysis, is appropriate for the high precision reconstruction of $W^\pm$ or Z^0 bosons, as it is designed to suppress muons from inflight decays and shows a low misidentification rate. Such a muon candidate must be a tracker muon as well as a global muon. The candidate track must have hits from at least six layers of the tracker, including at least one pixel hit, and it must be matched to at least two stations in the muon system. The fit of the global track must fulfill $\chi^2/N_{\text{dof}} < 10$ and include at least one hit from the muon system. The candidate must also be compatible with the primary vertex by requiring the impact parameters to be $|dxy| < 0.2\,\text{cm}$ and $|dz| < 0.5\,\text{cm}$.

The origin of a muon candidate can be found by applying an isolation criterion. Prompt muons have a higher isolation than muons from weak decays, which are often contained in jets. The isolation is measured relative to the muon p_T, summing up the energy deposited in the calorimeters and track momenta inside a cone $\Delta R = \sqrt{(\Delta\phi)^2 + (\Delta\eta)^2}$, around the muon axis and correcting for pileup contributions.

4.3.1.1 Muon Identification and Isolation Efficiency

The efficiency of the ID and Isolation criteria for muons is determined using the so called *tag-and-probe* method [1, 14]. The latter utilizes di-muon decays of Z^0 bosons, in which one muon candidate is assigned as the *tag* and needs to fulfill the trigger, ID, and isolation conditions and the *probe* is used to evaluate the efficiency under investigation. All tracks with $p_T^{\text{track}} > 20\,\text{GeV}$ in an event are tested as probes and events were more than one probe fulfills all selection criteria are rejected to reduce background contributions. The efficiency of the muon isolation is studied in relation to a probe which has passed a specific muon ID. Events are accepted when the invariant mass of the tag and probe combination falls into a predefined Z^0 boson mass window. The efficiencies are determined in data and MC separately and the correction factor $\rho = \epsilon_{\text{Data}}/\epsilon_{\text{MC}}$ is applied to MC samples to adjust the efficiencies to the ones observed in data. Figure 4.3 presents the efficiency of the muon tight ID and tight isolation as a function of the muon pseudorapidity, determined for the full 2016 pp-collision data. The reduced efficiency close to $|\eta| = 0.3$ is due to the geometric properties of the detector. It marks the border between the central muon wheel and the two neighbouring wheels and therefore contains less instrumentation.

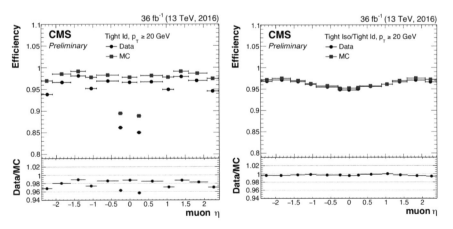

Fig. 4.3 Tight identification efficiency (left) and the tight isolation efficiency (right) as a function of the muon pseudorapidity, determined in data and $Z^0 \rightarrow \mu^+ \mu^-$ MC simulations [15]

4.3.1.2 Single Muon Triggers

The W + c analysis utilizes Single Muon triggers [1] for a preliminary event selection. The L1 muon trigger searches for hit patterns in the DT and CSC, consistent with a muon produced in the interaction region of the detector and the corresponding hit information is provided by the RPCs. For the HLT, simplified versions of the muon identification and muon isolation particle-flow algorithms are employed to reduce the computing time for the decision.

The p_T threshold for these triggers is derived using the tag-and-probe method, where the additional condition for the tag muon candidate to be matched to the HLT trigger is applied. The resulting efficiency curve rises sharply at the trigger threshold and reaches a plateau after a few GeV. Figure 4.4 presents the trigger efficiency of the SingleMuon trigger used in 13 TeV pp-collisions with a threshold of 24 GeV, determined from the pp-collision data recorded in 2016. Inefficiencies of a few percent at high p_T are due to the L1 trigger and the relative isolation conditions used in the HLT trigger, whereas inefficiencies in the high η regions are due to geometrical limitations of the detector affecting the L1 trigger efficiencies. The reduced efficiency close to $|\eta| = 0.3$ is again due to the borders between the central muon wheel and the two neighbouring wheels containing less instrumentation.

4.3.2 Missing Transverse Momentum

The momentum of particles that do not interact with the detector material, such as neutrinos, is approximated by the calculation of the missing transverse momentum $\vec{p}_T^{\text{miss}}$ [17] of an event. Due to momentum conservation, events containing such par-

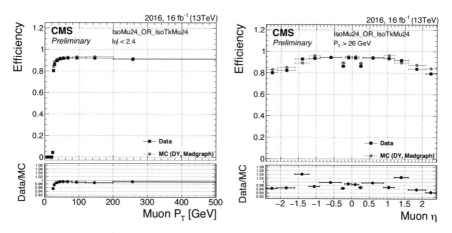

Fig. 4.4 Trigger efficiency of the SingleMuon trigger as a function of p_T^μ and η^μ in data and simulation [16]

Fig. 4.5 Illustration of the
measured momentum
imbalance in the CMS
detector [18]

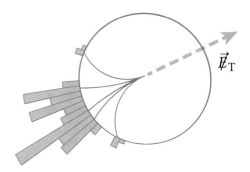

ticles show an imbalance of measured energy in the transverse plane of the detector. The concept is illustrated in Fig. 4.5. The missing transverse momentum is defined as the negative vector sum of the transverse momenta of all reconstructed particle-flow particles. Aside from the raw $\vec{p}_T^{\text{miss}}$, a corrected version including the jet-energy corrections (Eq. 4.1 term in brackets) is available:

$$\vec{p}_T^{\text{miss}} = - \sum_{i=1}^{N_{\text{particles}}} \vec{p}_T^i - \left[\sum_{j=i}^{N_{\text{PFjets}}} \left(p_{T,j}^{\text{corr}} - p_{T,j} \right) \right] . \tag{4.1}$$

Uncertainties in the determination of $\vec{p}_T^{\text{miss}}$ are caused by energy thresholds set in the calorimeters, non-linearities in their responses and inefficiencies in the reconstruction of tracks.

References

1. Sirunyan AM, et al (2018) Performance of the CMS muon detector and muon reconstruction with proton-proton collisions at ps = 13 TeV. JINST 13:P06015. https://doi.org/10.1088/1748-0221/13/06/P06015. arXiv: 1804.04528 [physics.ins-det]
2. Sirunyan AM, et al (2017) Particle-flow reconstruction and global event description with the CMS detector. JINST 12:P10003. https://doi.org/10.1088/1748-0221/12/10/P10003. arXiv: 1706.04965 [physics.ins-det]
3. Billoir P (1989) Progressive track recognition with a Kalman-like fitting procedure. Comput Phys Commun 57(1):390–394. ISSN: 0010-4655. https://doi.org/10.1016/0010-4655(89)90249-X. http://www.sciencedirect.com/science/article/pii/001046558990249X
4. Billoir P, Qian S (1990) Simultaneous pattern recognition and track fitting by the Kalman filtering method. Nucl Instrum Methods Phys Res Sect A: Accel, Spectrometers, Detect Assoc Equip 294(1):219–228. ISSN: 0168-9002. https://doi.org/10.1016/0168-9002(90)91835-Y. http://www.sciencedirect.com/science/article/pii/016890029091835Y
5. Mankel R (1997) A concurrent track evolution algorithm for pattern recognition in the HERA-B main tracking system. Nucl Instrum Methods Phys Res Sect A: Accel, Spectrometers, Detect Assoc Equip 395(2):169-184. ISSN: 0168-9002. https://doi.org/10.1016/S0168-9002(97)00705-5. http://www.sciencedirect.com/science/article/pii/S0168900297007055

6. Adam W et al (2006) Track Reconstruction in the CMS tracker. Technical report CMS-NOTE-2006-041. Geneva: CERN. https://cds.cern.ch/record/934067

7. Chatrchyan S et al (2014) Description and performance of track and primary vertex reconstruction with the CMS tracker. JINST 9:P10009. https://doi.org/10.1088/1748-0221/9/10/P10009. arXiv: 1405.6569 [physics.ins-det]

8. Miao T et al (2007) Beam position determination using tracks. Technical report CMSNOTE-2007-021. Geneva: CERN. https://cds.cern.ch/record/1061285

9. Chatrchyan S et al (2014) Alignment of the CMS tracker with LHC and cosmi ray data. JINST 9:P06009. https://doi.org/10.1088/1748-0221/9/06/P06009. arXiv: 1403.2286 [physics.ins-det]

10. Chatrchyan S et al (2010) Alignment of the CMS silicon tracker during commissioning with cosmic rays. JINST 5:T03009. https://doi.org/10.1088/1748-0221/5/03/T03009. arXiv: 0910.2505 [physics.ins-det]

11. Moser H-G (2009) Silicon detector systems in high energy physics. Prog Part Nucl Phys 63:186–237. https://doi.org/10.1016/j.ppnp.2008.12.002

12. Rose K (1998) Deterministic annealing for clustering, compression, classification, regression, and related optimization problems. Proc IEEE 86(11):2210-2239. ISSN: 0018-9219. https://doi.org/10.1109/5.726788

13. Frühwirth R, Waltenberger W, Vanlaer P (2007) Adaptive vertex fitting. Technical report CMS-NOTE-2007-008. https://cds.cern.ch/record/1027031

14. Chatrchyan S et al (2011) Measurement of the inclusive W and Z production cross sections in pp collisions at ps = 7 TeV. JHEP 10:132. https://doi.org/10.1007/JHEP10(2011)132. arXiv: 1107.4789 [hep-ex]

15. Muon Identification and Isolation efficiency on full 2016 dataset (2017). https://cds.cern.ch/record/2257968

16. Muon HLT performance on 2016 data (2017). https://cds.cern.ch/record/2297529

17. The CMS Collaboration (2016) Performance of missing energy reconstruction in 13 TeV pp collision data using the CMS detector. CMS Physics Analysis Summary CMS-PAS-JME-16-004. https://cds.cern.ch/record/2205284

18. Tai Sakuma for the CMS collaboration (2012) Missing ET schematic diagram. https://cms-docdb.cern.ch/cgi-bin/DocDB/ShowDocument?docid=12312

Chapter 5
Measurement of the Associated Production of W + Charm in $\sqrt{s} = 13$ TeV Proton–Proton Collisions

This chapter presents the measurement of the associated production of a $W^\pm$ boson and a charm quark in 13 TeV pp-collisions, using the data collected by the CMS collaboration in 2016, corresponding to an integrated luminosity of $35.7\,\text{fb}^{-1}$. It is the main subject of this thesis, together with the interpretation of the results in terms of the proton structure, presented in Chap. 6.

In this analysis, the $W^\pm$ bosons are selected by requiring the presence of a high-p_T muon and missing transverse momentum, indicating the presence of a neutrino. The charm quarks are tagged via the reconstruction of charmed mesons in their decay chain $c \to D^*(2010)^\pm \to D^0\pi^\pm_{\text{slow}} \to K^\mp\pi^\pm\pi^\pm_{\text{slow}}$, and no association of the $D^*(2010)^\pm$ meson to a jet is required. Due to the small mass difference between $D^*(2010)^\pm$ and D^0 particles, the pion originating from the $D^*(2010)^\pm$ decay receives very little energy and is therefore denoted as "slow". This process provides a clear experimental signature which can be reconstructed with good accuracy by the CMS detector. The fiducial cross section of $W + c$ is determined inclusively, and as a function of the pseudorapidity of the muon originating from the $W^\pm$ boson decay. The results of the measurement are compared to theoretical predictions, calculated with MCFM 6.8 [1–3] at next-to-leading order in combination with different PDF sets. Additionally, the cross section of $W + D^*(2010)^\pm$ production is measured within a selected fiducial phase space. The results will eventually be used to create a Rivet plugin [4], a tool used in the validation and tuning of MC event generators. However, this program requires a particle level measurement without restrictions on the origin of the particles to specific partons or production processes.

© The Editor(s) (if applicable) and The Author(s), under exclusive license
to Springer Nature Switzerland AG 2020
S. K. Pflitsch, *Associated Production of W + Charm in 13 TeV
Proton-Proton Collisions Measured with CMS and Determination of the Strange Quark
Content of the Proton*, Springer Theses,
https://doi.org/10.1007/978-3-030-52762-4_5

5.1 Signal Definition and Samples

Events containing the associated production of $W^+ + \bar{c}$ or $W^- + c$ are characterized by the presence of a W boson and a charm quark with opposite charge signs (OS). The dominant background contribution for this process is $W + c\bar{c}$, which has a similar event signature, with a $W^\pm$ boson produced from hard scattering and the charm quark originating from gluon splitting. Possible production mechanisms for $W + c\bar{c}$ are: $u + \bar{d} \rightarrow W^+ + g^* \rightarrow W^+ + c\bar{c}$ or $d + \bar{u} \rightarrow W^- + g^* \rightarrow W^- + c\bar{c}$.

However, in addition to the OS combination, $W + c\bar{c}$ events produce an equal amount of combinations, where the $W^\pm$ boson and the c quark have the same charge sign (SS). Therefore, the contribution from gluon splitting in the signal can largely be removed, using a data driven method based on the charge signs of the $W + c$ ($\bar{c}$) combinations. The details are covered in Sect. 5.2.3.

5.1.1 Data Samples

In this analysis, the data collected by the CMS experiment in $\sqrt{s} = 13\,\mathrm{TeV}$ pp collisions in 2016 is used. The performance of the detector is monitored during data taking and the reconstructed events are undergoing certification to ensure high quality for physics analysis [5]. After certification, the recorded events correspond to an integrated luminosity of $\mathcal{L} = 35.7\,\mathrm{fb}^{-1}$ [6]. In this analysis, single muon triggers are used for a preliminary event selection, as the $W^\pm$ bosons are reconstructed via their leptonic decay to a muon and a neutrino. A detailed list of the different run periods, the number of events and the certification file used, can be found in Appendix A.

5.1.2 Simulated Samples

The signal and background processes are simulated using different Monte Carlo (MC) generators to estimate the acceptance and efficiency of the CMS detector and to determine background contributions after the event and object selection. The reconstruction of the generated events is done via a simulation of the CMS detector based on the GEANT4 [7] program.

Additional pileup events are simulated on top of the hard scattering process, such that the average number of pileup interactions corresponds to the expected distribution in data. The remaining discrepancies in the simulation to the observed distribution of reconstructed primary vertices in data are adjusted by reweighting the *true number of interactions*, provided by the MC truth information, and a reference pileup distribution derived from data. The latter is determined by measuring the instantaneous luminosity per bunch crossing, the total integrated luminosity of the data taking period, and the total inelastic cross section of pp-collisions. Figure 5.1 presents the number of reconstructed primary interactions before and after the reweighting.

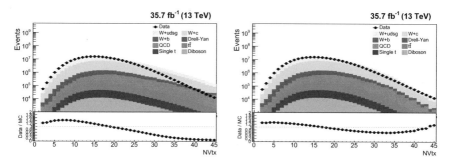

Fig. 5.1 Number of reconstructed primary vertices before (left) and after (right) reweighting. The data (filled circles) are compared to MC simulations of contributions from different processes (filled bands of different colour)

All of the simulated samples are normalized to the integrated luminosity of the data and use the NNPDF3.0nlo [8] PDF set to describe the proton structure. Furthermore, all generators are interfaced with PYTHIA (v8.2.12) [9] for parton shower and hadronization. The underlying event tune applied to all samples is CUETP8M1 [10], unless otherwise specified.

All $W^{\pm}$ + jets samples are generated using MADGRAPH5_aMC@NLO (v2.2.2) [11], with NLO matrix elements used in the calculations, and the FxFx [12] technique is applied for matching and merging. The factorization (μ_f) and renormalization (μ_r) scales are set to $\mu_f^2 = \mu_r^2 = M_W^2 + p_{T,W}^2$. The signal is simulated using an inclusive $W^{\pm}$ + jets sample and applying a filter to select $W + c$ and $W + c\bar{c}$ events. The filter requires the presence of least one muon with $p_T^{\mu} > 20$ GeV and $|\eta^{\mu}| < 2.4$, as well as at least one $D^*(2010)^{\pm}$ meson in the event. Furthermore, an inclusive $W^{\pm}$ + jets sample without an applied filter is generated to estimate background contributions to the event selection.

The following definitions are applied to all $W^{\pm}$ + jets samples at the generator level. An event is considered to be a $W + c$ candidate if it contains at least one c quark in the final state. There are no separate event categories for $W + c$ and $W + c\bar{c}$, but all c quarks in an event are marked as originating either from $W + c$ or from gluon splitting. In the case of an odd number of c quarks in an event, the c quark with the highest p_T and a charge opposite to that of the $W^{\pm}$ boson is considered to originate from a $W + c$ process, whereas the other c quarks in the event are marked as gluon splitting. In the case of an even number of c quarks, all are marked as coming from gluon splitting. Events containing both c and b quarks are considered to be $W + c$ events, due to $W + c$ being of more importance in this analysis, while the events containing no c but at least one b quark are classified as $W^{\pm} + b(\bar{b})$. Anything that falls in neither of these categories is assigned to $W^{\pm} + udsg$.

In the $W + D^*$ measurement, no constrains on the origin of the $D^*(2010)^{\pm}$ mesons are applied at the generator level. Therefore, any contributions from B meson decays and other hadrons, though only a few pb, are included as signal for this part of the measurement.

Further processes are considered due to their contributions to the background, which are simulated as follows: Events originating from Drell–Yan processes are generated with MADGRAPH5_aMC@NLO (v2.2.2) and the scales are set to $\mu_f^2 = \mu_f^2 = M_Z^2 + p_{T,Z}^2$. Top quark-antiquark pairs are generated using the POWHEG (v2.0) [13] generator and the CUETP8M2T4 [14] underlying event tune is applied. Single top quark events are simulated using POWHEG (v1.0) [15] for the tW production channel and POWHEG (v2.0) [16, 17] to simulate the t-channel and s-channel production. Events containing multiple $W^\pm$ or Z^0 bosons (WW, WZ, ZZ) are generated using PYTHIA (v8.2.12) for all steps of the event simulation. Contributions from muons, originating from inflight decays of charm or bottom hadrons are generated using PYTHIA (v8.2.12) as well.

5.2 Event Selection

The associated production of $W + c$ is investigated via the leptonic decay of $W^\pm$ bosons into a muon and a neutrino ($W^\pm \rightarrow \mu \bar{\nu}_\mu$) and a charm quark hadronizing to a $D^*(2010)^\pm$ meson. To ensure the presence of a high-p_T muon, an event is required to fulfill the trigger conditions of the single muon trigger (*IsoMu24* or *IsoTkMu24*) to be selected for further analysis (see Sect. 4.3.1.2). The p_T threshold for these triggers is derived from an efficiency curve with approximately 50% efficiency at 24 GeV, reconstructed with the HLT algorithms.

The charm quark is identified by the reconstruction a $D^*(2010)^\pm$ meson, the lowest excited state of the $D^\pm$ particle, with a mass of 2010.26 ± 0.05 MeV [18]. The quark content of this charmed meson is $c\bar{d}$ for a $D^*(2010)^+$ and $\bar{c}d$ for $D^*(2010)^+$ and has a decay width of $\Gamma = 83.4 \pm 1.8$ keV. With a spin of 1, the $D^*(2010)^\pm$ is a vector meson. The particle decays into a D^0 and a charged pion $D^*(2010)^\pm \rightarrow D^0\pi^\pm$ and the D^0 subsequently decays to a kaon and pion of opposite charge, which is considered the cleanest event signature.

5.2.1 W Boson Selection

Events containing a $W^\pm$ boson decay are identified by the presence of a high-p_T isolated muon and missing transverse momentum $\vec{p}_T^{\text{miss}}$, indicating the presence of a neutrino. The muons and $\vec{p}_T^{\text{miss}}$ contained in an event are reconstructed via the particle-flow algorithm.

Muon candidates are required to have a transverse momentum above $p_T^\mu > 26$ GeV and must fulfill the CMS *tight identification* criteria (see Sect. 4.3.1). To suppress the contamination from muons contained in jets, an isolation requirement is imposed as well:

$$I = \left[\sum^{CH} p_T + \max\left(0., \sum^{NH} p_T + \sum^{EM} p_T - 0.5 \sum^{PU} p_T \right) \right] \Big/ p_T^\mu \leq 0.15 , \quad (5.1)$$

It runs over the p_T sum of particle-flow (PF) candidates for charged hadrons (CH), neutral hadrons (NH), photons (EM) and charged particles from pileup (PU) inside a cone of radius $\Delta R \leq 0.4$. The factor 0.5 corresponds to the typical ratio of neutral to charged particles, as measured in jet production [19].

Events in which more than one muon candidate fulfills all the above criteria are rejected in order to suppress background from Drell–Yan processes. Corrections are applied to the simulated samples to adjust the trigger, isolation, identification, and tracking efficiencies to the observed data. These correction factors are determined by performing dedicated tag-and-probe studies [20] (see Sect. 4.3.1.1).

The presence of a neutrino in an event is assured by imposing a requirement on the transverse mass (M_T), which is defined as a product of p_T^μ and $\vec{p}_T^{\,miss}$:

$$M_T \equiv \sqrt{2 \cdot p_T^\mu \cdot \vec{p}_T^{\,miss} \, (1 - \cos(\phi_\mu - \phi_{\vec{p}_T^{\,miss}}))}. \quad (5.2)$$

In this analysis, $M_T > 50\,\text{GeV}$ is required, resulting in a significant suppression of muons from hadron decays. This particular background is referred to as *QCD background* throughout the chapter.

The p_T, η and isolation distribution of the muon candidates, fulfilling all event selection criteria, are presented in Fig. 5.2, and the distributions for $\vec{p}_T^{\,miss}$ and M_T are presented in Fig. 5.3. Fairly good agreement between the data and Monte Carlo predictions is observed for all distributions.

5.2.2 Selection of D⁰ and D*(2010)± Candidates

The D*(2010)⁻ candidates are identified by their decays $D^*(2010)^\pm \to D^0 + \pi_{slow}^\pm \to K^\mp \pi^\pm \pi_{slow}^\pm$, which has a branching fraction of $2.66 \pm 0.03\,\%$ [18]. As all of the final decay products are long lived charged particles, they can be reconstructed as tracks in the detector. This particular channel provides a clear signal, as the D*(2010)± meson is reconstructed via the mass difference between the D⁰ and D*(2010)± candidates, instead of directly reconstructing the invariant mass distribution of the D*(2010)±. By using this method, the uncertainties associated with the tracks associated with a D⁰ candidate are largely cancelled and the resolution of $\Delta m(D^0, D^*)$ depends almost solely on the uncertainties of the third track.

The D⁰ candidates are constructed by combining two oppositely charged tracks with a transverse momentum of $p_T^{track} > 1\,\text{GeV}$, assuming the masses of a $K^\mp$ and a $\pi^\pm$. The distributions of p_T, η and ϕ of the candidates are presented in Fig. 5.4 and the relative track resolution is shown in Fig. 5.5 as a function of p_T^{track} and η^{track}. The latter shows that the track resolution achieved in data is well reproduced in the simulation. Additional selection criteria of the analysis are: The impact parameters of both tracks

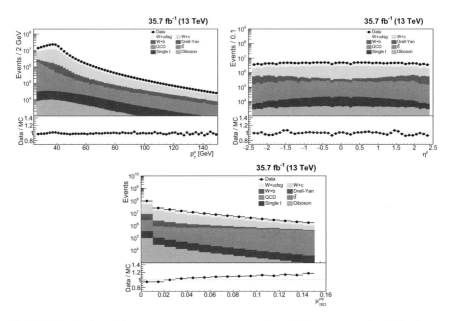

Fig. 5.2 Distribution of the muon transverse momentum (upper left), muon pseudorapidity (upper right) and muon isolation (lower plot). The data (filled circles) are compared to the MC simulation of contributions from different processes (filled bands of different color)

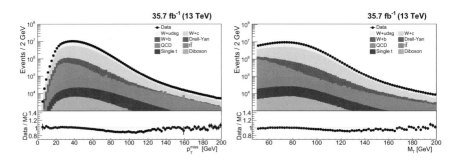

Fig. 5.3 Distributions of missing transverse momentum (left) and of transverse mass (right). The data (filled circles) are compared to the MC simulation of contributions of different processes (filled bands of different color)

must not be separated by more than 0.1 cm in the xy-plane and the z-coordinate, and they must originate at a fitted secondary vertex, determined by the adaptive vertex fitter [21]. Figure 5.6 presents the separation of the $K^{\mp}$ and $\pi^{\pm}$ candidates in the xy-plane and the z-coordinate. Only combinations with a reconstructed mass differing from the D^0-mass (1864.86 ± 0.13 MeV [18]) by less than 35 MeV are selected. Good agreement between the data and MC predictions is observed for all distributions.

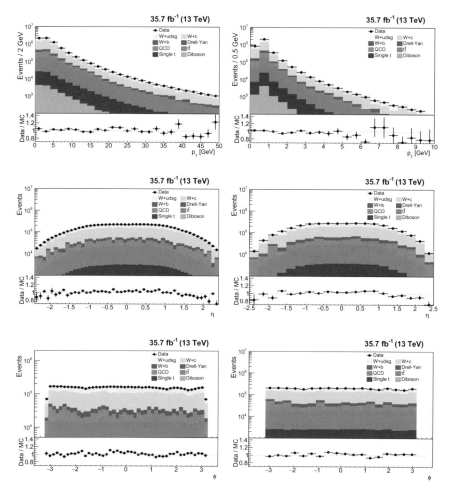

Fig. 5.4 Track p_T (upper), η (middle) and ϕ (lower) distributions for the $K^\mp$ and $\pi^\pm$ (left) and the $\pi^\pm_{slow}$ candidates in the events fulfilling all selection criteria. The data (filled circles) are compared to the MC simulation of contributions from different processes (filled bands of different color)

Both tracks of the D^0 candidate are further combined with a third track, which has a charge opposite to the $K^\mp$ candidate, and is presumed to be the $\pi^\pm_{slow}$. The track is required to have a minimum of $p_T^{track} > 0.35$ GeV, and it must be found in a cone of $\Delta R < 0.15$ around the D^0 axis.

As the CMS detector is designed to be most efficient in the reconstructino of high p_T particles, it is important to investigate whether the reconstructed p_T^{track} and resolution of low momentum tracks, which make up the majority of the $\pi^\pm_{slow}$ candidates, is well reproduced in the simulation. The cross section for pion nucleon interactions increases rapidly for pions with energies below 0.7 GeV [22], leading to a reduced reconstruction efficiency. At the same time, tracks with low momentum

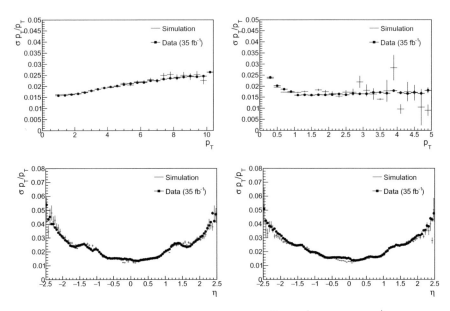

Fig. 5.5 Relative transverse momentum resolution of the $K^{\mp}$ and $\pi^{\pm}$ (left) and the $\pi_{slow}^{\pm}$ candidates (right) as a function of p_T^{track} (upper) and η^{track} (lower) in the events fulfilling all selection criteria

are prone to spiral in the magnetic field, thereby producing a large number of hits in the detector. Due to this, the probability of adding uncorrelated hits to such a track candidate are high, which increases the fake rate and resolution uncertainty. The p_T, η and ϕ distributions and the relative track resolution are presented separately for the $\pi_{slow}^{\pm}$ candidates in Figs. 5.4 and 5.5. The relative track resolution is well reproduced by the simulation even though statistical fluctuations occur for tracks above $p_T^{track} > 3$ GeV. In addition, the impact parameter of the $\pi_{slow}^{\pm}$ candidate must not be displaced from the D^0 vertex by more than 0.1 cm in the xy-plane or the z-coordinate. The distribution of these observables are presented in Fig. 5.6 as well.

The combinatorial background is reduced by requiring the $D^*(2010)^{\pm}$ transverse momentum to be $p_T^{D^*} > 5$ GeV and by imposing an isolation requirement $p_T^{D^*} / \sum p_T^{track} > 0.2$. Here, $\sum p_T^{track}$ is the sum of transverse momenta of tracks in a cone of $\Delta R < 0.4$ around the direction of the $D^*(2010)^{\pm}$ momentum. The contribution of $D^*(2010)^{\pm}$ mesons produced in pileup events is suppressed by rejecting candidates with a distance of $\Delta z > 0.2$ cm between the muon associated with the $W^{\pm}$ candidate and the slow pion. Both distributions are presented in Fig. 5.7 and the p_T, η and ϕ distributions of the D^0 and $D^*(2010)^{\pm}$ candidates is shown in Fig. 5.8. Additional investigations that had no significant impact on the $D^*(2010)^{\pm}$ selection, can be found in Appendix B.

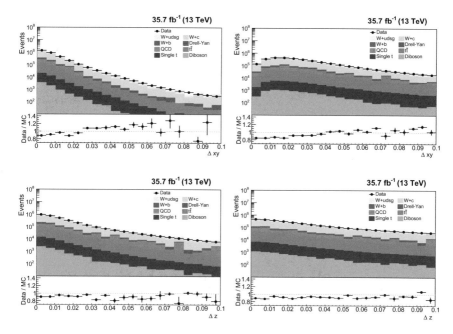

Fig. 5.6 Separation in the xy-plane (upper) and z-coordinate (lower) between the $K^{\mp}$ and $\pi^{\pm}$ candidates (left) and between the D^0 candidate and slow pion candidate (right), fulfilling all selection criteria. The data (filled circles) are compared to the MC simulation of contributions from different processes (filled bands of different color)

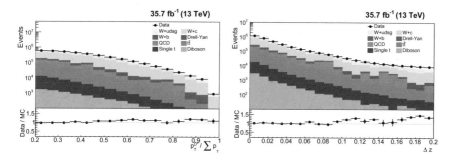

Fig. 5.7 Transverse momentum fraction of the D^* candidate (left) and z-separation of the $\pi^{\pm}_{slow}$ and μ, originating from the decay of the $W^{\pm}$ boson candidate. The data (filled circles) are compared to the MC simulation of contributions from different processes (filled bands of different color)

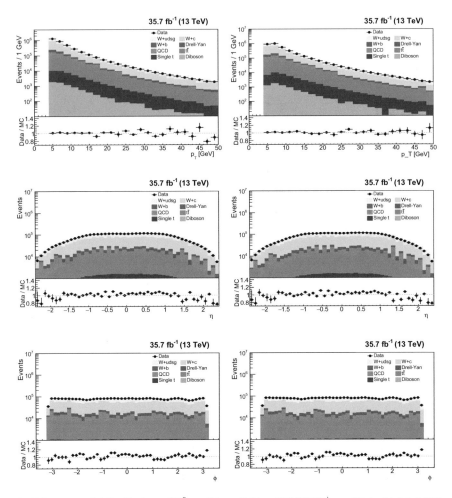

Fig. 5.8 p_T, η and ϕ distributions of D^0 candidates (left) and $D^*(2010)^{\pm}$ candidates (right), fulfilling all selection criteria. The data (filled circles) are compared to the MC simulation of contributions from different processes (filled bands of different color)

5.2.2.1 Subtraction of Combinatorial Background

The combinatorial background is significantly suppressed by using a data-driven method [23]. For this purpose, a set of *fake* D^0 candidates are constructed from two track combinations, in which both tracks have the same charge (wrong-charge), while all other selection criteria remain unchanged. The resulting distribution mimics the shape of the light flavour background for D^0 candidates (right-charge). Differences in the normalization of both distributions are related to contributions from the decays of D^0 mesons that do not originate from the decay of a $D^*(2010)^{\pm}$. To correct for the difference in the distributions, the ratio of the right-charge and wrong-charge is

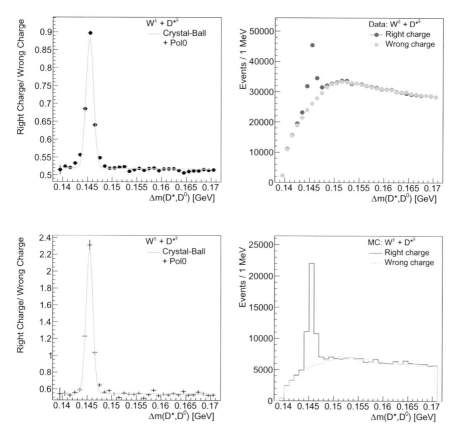

Fig. 5.9 Ratio between the $\Delta m(\mathrm{D}^0, \mathrm{D}^*)$ distributions of the right-charge and wrong-charge W + D* candidates (left), fitted with a crystal-ball function plus a constant factor. The right-charge and wrong-charge distributions (right) are presented after the normalization of the wrong-charge distribution, using the constant factor of the fitted function. Data (upper) and simulation (lower) are presented separately

fitted with a Crystal Ball function [24] plus a constant factor, thereby including the signal region $(0.14 < \Delta m(\mathrm{D}^0, \mathrm{D}^*) < 0.15)$ in the fit.

The wrong-charge distribution is then scaled with the constant part of the ratio-function and subtracted from the right-charge distribution. The ratio between the right-charge and wrong-charge distribution is presented for W + D*(2010)$^{\pm}$ candidate events in Fig. 5.9, together with the right-charge and wrong-charge distributions after normalization.

An analogue procedure is employed to subtract the combinatorial background of the $\mathrm{K}^{\mp} \pi^{\pm}$ invariant mass distribution. In this case, the D^0 candidates are selected if the mass difference to the corresponding D*(2010)$^{\pm}$ candidate is found in a window of $|\Delta m(\mathrm{D}^0, \mathrm{D}^*) - 0.1454| < 0.001\,\mathrm{GeV}$. Here, the ratio of the right-charge and wrong-charge candidates is fit with a second order polynomial, but excluding the

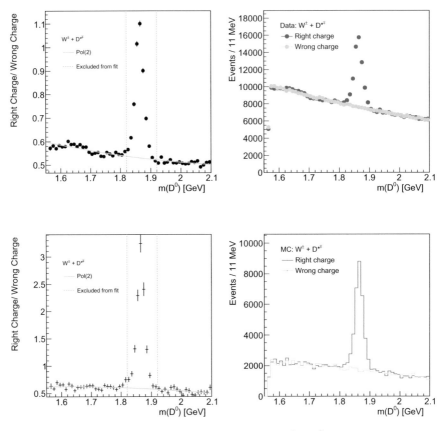

Fig. 5.10 Ratio between the right-charge and wrong-charge $W^\pm + D^0$ candidates (left), selected in a window of $|\Delta m(D^0, D^*) - 0.1454| < 0.001$ GeV. The distribution is fit with a second order polynomial, excluding the region of $1.82 < m(K^\mp, \pi^\pm) < 1.92$ GeV. The right-charge and wrong-charge distributions after normalization of the wrong-charge candidates are presented (right) separately for data (upper) and simulation (lower)

region of $1.82 < m(K^\mp, \pi^\pm) < 1.92$ GeV, where the D^0 peak is located. The wrong-charge candidates are then rescaled, using the fitted function. Figure 5.10 presents the ratio between the right-charge and wrong-charge candidates and both individual distributions after the rescaling of the wrong-charge distribution.

5.2.3   Selection of W + c Candidates

An event is selected as a W + c signal if it contains a $W^\mp$ boson and a $D^*(2010)^\pm$ candidate fulfilling all selection criteria. The candidate events are assigned to one of two categories: with $W^\pm + D^*(2010)^\pm$ combinations falling into the same sign (SS)

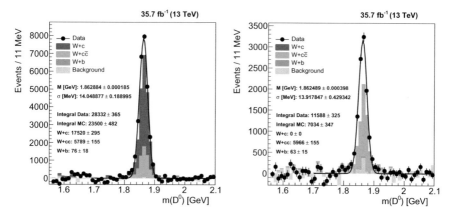

Fig. 5.11 Distribution of the invariant mass of D^0 candidates, in the range $|\Delta m(D^0, D^*) - 0.1454| < 0.001$ GeV, assigned to the OS category (left) or SS category (right). The combinatorial background is subtracted. The data (filled circles) are compared to MC simulation with contributions from different processes shown as histograms of different colours

category, and $W^{\pm} + D^*(2010)^{\mp}$ combinations into to the opposite sign (OS) category. Figure 5.12 presents the $\Delta m(D^0, D^*)$ distributions for events falling into the OS and SS categories, with the combinatorial background of the distributions already removed. A clear peak around the expected value of 145.4257 ± 0.0017 MeV [18] is observed. The corresponding invariant mass distribution of the D^0 candidates, selected in a $\Delta m(D^0, D^*)$ window of ± 1 MeV, and separated into the OS and SS categories, is presented in Fig. 5.11. A clear peak around the expected value of 1864.86 ± 0.13 MeV is observed.

The OS combinations are composed of signal events and contributions from $W + c\bar{c}$ and $W + b\bar{b}$, whereas the SS combinations are comprised of contributions from background processes, only. At this stage, neither the OS nor the SS distributions show a good agreement between data and simulation. This is due to an insufficient description of gluon splitting in the used Monte Carlo simulations. However, since the number of $W + c\bar{c}$ and $W + b\bar{b}$ candidates in both categories is equal within uncertainties, subtracting the SS events from the OS events removes the gluon splitting background. The contributions from other background sources, such as $t\bar{t}$ and single top quark production, are negligible.

After the subtraction of the SS contributions in the OS distribution, the number of reconstructed $D^*(2010)^{\pm}$ mesons corresponds to the number of $W + c$ events. The observed reconstructed mass difference $\Delta m(D^0, D^*)$ is presented in Fig. 5.13, along with the invariant mass of $K^{\pm} \pi^{\mp}$ candidates, selected in a $\Delta m(D^0, D^*)$ window of ± 1 MeV. The remaining $W + c\bar{c}$ and $W + b\bar{b}$ background is negligible, and the number of $D^*(2010)^{\pm}$ mesons is determined by counting the number of candidates in a window of $144 < \Delta m(D^0, D^*) < 147$ MeV. Alternately, two different functions are fit to the distributions, and their integral over the same mass window is used to estimate the systematic uncertainties associated with the chosen method.

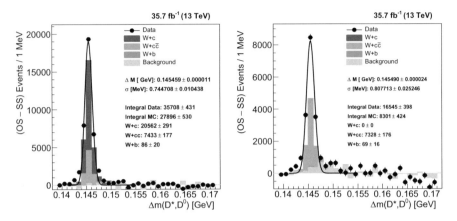

Fig. 5.12 Distribution of the reconstructed mass difference $\Delta m(D^0, D^*)$ for candidates assigned to the OS category (left) or SS category (right). The combinatorial background is subtracted. The data (filled circles) are compared to MC simulation with contributions from different processes shown as histograms of different colours

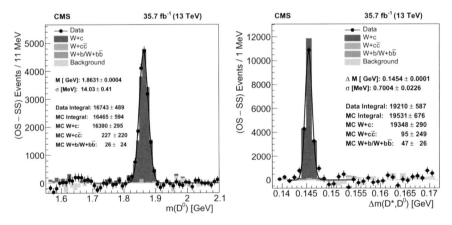

Fig. 5.13 Distributions of the reconstructed invariant mass of $K^{\pm}\pi^{\mp}$ candidates (left) in the range $|\Delta m(D^0, D^*) - 0.1454| < 0.001$ GeV, and the reconstructed mass difference $\Delta m(D^0, D^*)$ (right). The SS combinations are subtracted. The data (filled circles) are compared to MC simulation with contributions from different processes shown as histograms of different shades

5.3   Measurement of the W + Charm Cross Section

The kinematic region of the fiducial $W + c$ measurement is defined by the requirements of the transverse momentum and the pseudorapidity of the muon, and the transverse momentum of the charm quark. It corresponds to $p_T^{\mu} > 26$ GeV, $|\eta^{\mu}| < 2.4$, and $p_T^c > 5$ GeV, taking the constraints on the reconstruction of the $W^{\pm}$ and $D^*(2010)^{\pm}$ candidate into account.

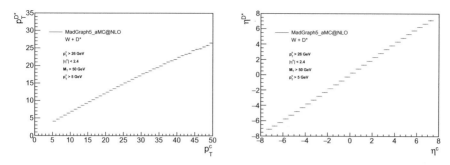

Fig. 5.14 Average $p_T^{D^*}$ as a function of p_T^c (left) and average η^{D^*}, as a function of η^c

The transverse momentum of the charm quark is determined by extrapolating from the kinematic phase space of the $D^*(2010)^\pm$ meson to the phase space of the charm quark, using the signal sample. The constraint on the charm quark is a result of the requirement on the transverse momentum of the D^* candidates, whereas the other kinematic properties of the $D^*(2010)^\pm$ mesons are integrated over at the generator level. The correlation between the kinematic properties of the $D^*(2010)^\pm$ mesons and that of the corresponding charm quark are investigated using MC, as is illustrated in Fig. 5.14, which presents the average transverse momentum of the $D^*(2010)^\pm$ mesons as a function of the charm quark p_T and the average η^{D^*} as a function of η^c. A linear correlation between the kinematics of $D^*(2010)^\pm$ meson and the mother charm quark is observed.

A comparison of the p_T^μ and p_T^c differential distribution between the signal simulation and NLO W + c calculations, using MCFM 6.8 [1–3] with NNPDF3.0nlo is presented in Fig. 5.15. The signal simulation corresponds to an inclusive W + D* sample and is therefore scaled with the inverse fragmentation probability $[Br(c \to D^*) = 0.2429]^{-1}$ [25] for the comparison to the W + c prediction. Moreover, the MC signal samples is not a pure pQCD calculation but missing higher order corrections are approximated by using a parton shower which dominantly contributes to the low p_T region of the distribution. Therefore, in addition to the inclusive spectrum, the distributions of p_T^c are presented separately for generated particles associated with the hard scattering process and those handled by the parton shower. It is observed that the MC sample is reproducing the shape of the distributions obtained with the fixed order prediction in the high-p_T region, whereas the low p_T region, dominated by the parton shower shows some expected differences. Due to a generally good agreement of the charm quark kinematics between the MC and the NLO calculation in the high p_T^c region, and the correlation between p_T^c and $p_T^{D^*}$, it is considered safe to translate the fiducial cut of $p_T^{D^*} > 5$ GeV into $p_T^c > 5$ GeV.

The fiducial cross section of W + c production is determined as follows:

$$\sigma(W + c) = \frac{N_{sel}\, S}{\mathcal{L}_{int}\, \mathcal{B}(c \to D^*; D^* \to D^0 + \pi_{slow}^\pm; D^0 \to K^\mp \pi^\pm)\, \mathcal{C}}, \tag{5.3}$$

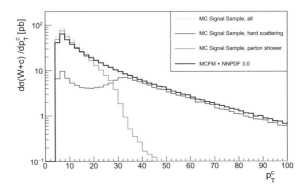

Fig. 5.15 Fiducial differential cross section of W + c as a function of p_T^c. The distributions of the MADGRAPH5_aMC@NLO signal MC sample, corrected for the branching ratio $\mathcal{B}(c \rightarrow D^*(2010)^{\pm}) = \times 0.2429$, are compared to the calculations done with MCFM in combination with the NNPDF3.0nlo PDF set

with N_{sel} as the number of events, selected in the $\Delta m(D^0, D^*)$ distribution after the subtraction of the SS events from the OS ones ($OS - SS$). The signal fraction $\mathcal{S}$ is defined as the ratio of the number of reconstructed W + D* candidates originating from W + c to the number of all reconstructed D*. It is determined from the MC simulation and accounts for any residual background contributions from gluon splitting or combinatorial background, caused by statistical fluctuations. The integrated luminosity is denoted by $\mathcal{L}_{int}$. The branching fraction $\mathcal{B}$ is a combination of the probability of a charm quark hadronizing to a $D^*(2010)^{\pm}$ meson $\mathcal{B}(c \rightarrow D^*) = 0.2429 \pm 0.0049$ [25], and the branching fractions of the decay channels $\mathcal{B}(D^*(2010)^{\pm} \rightarrow K^{\pm} + \pi^{\mp} + \pi_{slow}^{\pm}) = 0.0266 \pm 0.0003$ [18], studied in this measurement. The correction factor $\mathcal{C}$ accounts for the acceptance and efficiency of the detector. It is determined using the MC simulation and is defined as the ratio of the number of reconstructed W + D* candidates to the number of generated W + D* originating from W + c events that fulfill the fiducial requirements.

As a parton level measurement, the W + c cross section depends to a large extent on the MC simulation for the extrapolation to unmeasured phase space and therefore has to take the uncertainties, associated with the method, into account. To reduce the amount of extrapolation in the corresponding particle level measurement, the W + D* cross section is determined in the fiducial phase space of the detector level measurement, $p_T^{\mu} > 26$ GeV, $|\eta^{\mu}| < 2.4$, $|\eta^{D^*}| < 2.4$ and $p_T^{D^*} > 5$ GeV. Therefore, Eq. (5.3) is modified as follows:

The branching fraction $\mathcal{B}$ is reduced to $\mathcal{B} = \mathcal{B}(D^* \rightarrow K^{\pm} + \pi^{\mp} + \pi_{slow}^{\pm})$, as the origin of the $D^*(2010)^{\pm}$ meson is not restricted to charm quarks for this part of the analysis. The factor $\mathcal{C}$ corresponds to the ratio between the number of reconstructed W + D* events and the number of generated W + D* candidates fulfilling the fiducial requirements, after OS − SS subtraction.

The cross sections of both, the W + c measurement, as well as the W + D*(2010)$^{\pm}$ measurement, are determined inclusively and also in five bins of the absolute pseu-

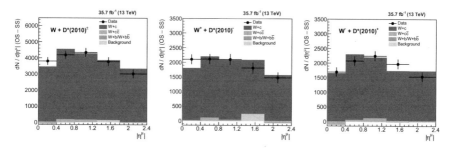

Fig. 5.16 Number of events after OS − SS subtraction for data (filled circles) and MC simulation (filled histograms) as a function of $|\eta^\mu|$. The distributions are presented for the combination of W^+ and W^- (left), and for the individual contributions of W^+ (middle) and W^- (right)

dorapidity $|\eta^\mu|$ of the muon originating from the $W^\pm$ boson decay. The number of signal (OS − SS) events in the different ranges of $|\eta^\mu|$ under study, is presented in Fig. 5.16 for $W + D^*$, $W^+ + D^*(2010)^-$ and $W^- + D^*(2010)^+$. Good agreement between the data and MC simulation within the statistical uncertainties is observed.

5.3.1 Systematic Uncertainties

The influence of systematic uncertainties on the measured cross sections is determined by varying the efficiencies and assumptions within their respective uncertainties. The resulting shifts in the cross section, with respect to the central value, are taken as the contribution to the systematic uncertainty of the measurement. The largest shift in each direction is accounted for in the overall uncertainty. The various sources of the systematic uncertainties, considered for the $W + c$ production cross section, are listed in Table 5.1. The contributions to the inclusive and the differential measurements are listed separately.

Due to the low momentum of the $\pi^\pm_{slow}$ candidates, it is necessary to use a special data format which contains all of the tracking information for low-momentum tracks ($p_T^{track} < 0.9\,\text{GeV}$), which was not provided in the standard data format at the time of this analysis. However, simulations in this special data format are only available in a limited capacity, which leads to large MC statistics uncertainties, especially in the differential part of the analysis.

5.3.1.1 Luminosity

Uncertainties associated with the integrated luminosity measurement are estimated as 2.5% [6].

Table 5.1 Systematic uncertainties (%) in the inclusive and differential W + c cross section measurement in the fiducial region of the analysis. The total uncertainty corresponds to the sum of the individual contributions in quadrature. The contributions listed in the top part of the table cancel in the ratio

Pseudorapidity $[\lvert\eta^{\mu}\rvert]$	[0, 2.4]	[0, 0.4]	[0.4, 0.8]	[0.8, 1.3]	[1.3, 1.8]	[1.8, 2.4]
Luminosity	±2.5	±2.5	±2.5	±2.5	±2.5	±2.5
Tracking	±2.3	±2.3	±2.3	±2.3	±2.3	±2.3
Branching	±2.4	±2.4	±2.4	±2.4	±2.4	±2.4
Muons	±1.2	±1.2	±1.2	±1.2	±1.2	±1.2
N_{sel} determination	±1.5	±1.5	±1.5	±1.5	±1.5	±1.5
D* kinematics	±0.5	±0.5	±0.5	±0.5	±0.5	±0.5
Background normalization	± 0.5	+0.9/−0.8	+1.9/−0.8	+1.4/−0.5	+0.8/−1.0	0.0/−0.6
$\vec{p}_T^{miss}$	+0.7/−0.9	+0.4/−1.2	+1.3/−0.3	+1.1/−1.0	0.0/−2.6	0.0/+1.5
Pileup	+2.0/−1.9	+0.4/−0.5	+2.9/−3.0	+2.0/−1.9	+4.6/−5.1	+2.7/−2.6
Secondary vertex	−1.1	+1.3	−1.2	−1.5	−2.7	−2.5
PDF	±1.2	±1.3	±0.9	±1.4	±1.5	±1.7
Fragmentation	+3.9/−3.2	+3.4/−1.8	+7.4/−5.2	+3.3/−3.0	+2.2/−1.2	+7.4/−5.7
MC statistics	+3.6/−3.3	+8.8/−7.5	+9.0/−11.9	+7.9/−6.8	+9.8/−14.1	+10.1/−8.5
Total	+7.5/−7.0	+10.7/−9.3	+13.2/−14.2	+10.1/−9.3	+12.7/−16.2	+13.8/−12.1

5.3.1.2 Tracking Efficiency

The uncertainty in the tracking efficiency is determined using the method described in Ref. [26]. It exploits the ratio between the four body ($D^0 \to K^{\mp}\pi^{\pm}\pi^{\mp}\pi^{\pm}$) and two body decay ($D^0 \to K^{\mp}\pi^{\pm}$) of the neutral charm meson, originating from the decay of a D* via $D^* \to D^0 + \pi^{\pm}_{slow}$. The $D^*(2010)^{\pm}$ mesons are reconstructed using $\Delta m(D^0, D^*)$ and the light flavour background is estimated by a combined fit of the peak and the light flavour background. The number of reconstructed $D^*(2010)^{\pm}$ depends on the branching ratios and the reconstruction efficiency of the two body and four body decays:

$$\frac{N_{K3\pi}}{N_{K\pi}} = \frac{Br(D^0 \to K^{\mp}\pi^{\pm}\pi^{\mp}\pi^{\pm})}{Br(D^0 \to K^{\mp}\pi^{\pm})} \cdot \frac{\epsilon_{K3\pi}}{\epsilon_{K\pi}} = \mathcal{R} \cdot \frac{\epsilon_{K3\pi}}{\epsilon_{K\pi}} \qquad (5.4)$$

As the difference between both decays are two charged pions in the final state, the hadronic reconstruction efficiencies of data and simulation follow the relation:

$$\frac{N_{K3\pi}}{N_{K\pi}} \propto \epsilon_{Data}^2 \quad , \quad \frac{\epsilon_{K\pi}}{\epsilon_{K3\pi}} \propto \epsilon_{MC}^2 \qquad (5.5)$$

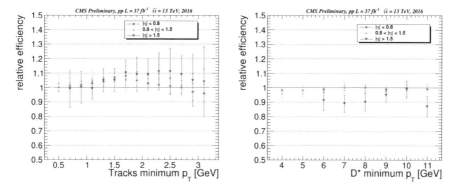

Fig. 5.17 Ratio of tracking efficiency between data and simulation as a function of the minimum p_T^{track} (left) and minimum $p_T^{D^*}$ (right). Different η regions of the tracker are represented by symbols of different colours [27]

Therefore, the relative tracking efficiency is calculated as follows, with $\mathcal{R}_{\text{PDG}}$ as the ratio of the two body and four body decays, taken from PDG.

$$\epsilon_{\text{rel}} = \frac{\epsilon_{\text{Data}}}{\epsilon_{\text{MC}}} = \sqrt{\frac{\mathcal{R}}{\mathcal{R}_{\text{PDG}}}}, \qquad (5.6)$$

The relative efficiency between data and simulation is presented in Fig. 5.17 as a function of the minimum track p_T and as a function of the minimum p_T of the $D^*(2010)^\pm$ candidates. Both distributions show that the tracking efficiency observed in data is well reproduced by the simulation for low p_T tracks.

The uncertainty associated with ϵ_{rel} is estimated by varying the kinematic range of the decay particles and the fit templates used to determine the number of reconstructed $D^*(2010)^\pm$ mesons. The resulting uncertainty in the tracking efficiency is 2.3% for the 2016 data.

5.3.1.3 Branching Fractions

The uncertainty related to the branching fractions is a combination of the uncertainties of the $D^* \rightarrow D^0 + \pi_{\text{slow}}^\pm$ (0.7%) and the $D^0 \rightarrow K^\pm \pi^\pm$ (1.0%) branching ratios, as they are listed in the PDG report of 2018 [18], and the probability of charm quarks hadronizing to a $D^*(2010)^\pm$ meson, which has been determined at 2% [25] accuracy. The latter is determined in a fit to the combined results sensitive to charm fragmentation, as measured by different experiments. The values are obtained via χ^2 minimization in a fit to the observables of interest, with the uncertainties included by using the covariance matrix representation described in Sect. 2.4.2. The corresponding uncertainties are evaluated using the Hessian method with a tolerance criterion of $\Delta\chi^2 = 1$. The individual contributions are added in quadrature, resulting in an uncertainty of 2.4%.

5.3.1.4 Muon Corrections

The uncertainties associated with the muon correction factors for *tight ID, tight isolation, trigger* and *tracking*, are estimated by varying the conditions used in the corresponding tag-and-probe analyses [20]. This includes an analysis of the background contamination by using variations on the p_T^μ requirements or the isolation for the tag. Furthermore, the requirement that only a single probe is associated with a tag is removed to investigate the impact on the efficiencies. However, the largest uncertainty originates from the assumptions on the signal and background models, used to fit the data, therefore, alternative fit functions are used to estimate the impact on the correction factors.

The muon systematic uncertainties of the 2016 data are 1% each for for the muon identification and isolation, and 0.5% for the trigger and tracking corrections. These are added in quadrature and the resulting uncertainty for muons is 1.2%, which is referred to as the 'muon uncertainty'.

5.3.1.5 N_{sel} Determination

The uncertainty in the determination of N_{sel} is estimated by using a Gaussian or Crystal Ball fit [24] as an alternative to counting the number of events in the final selection. The window in which the functions are integrated remains unchanged ($144 < \Delta m(D^0, D^*) < 147\,\text{MeV}$) and the resulting uncertainty corresponds to 1.5%.

5.3.1.6 $D^*(2010)^\pm$ Kinematics

Uncertainties in the modelling of kinematic observables of the generated D^* meson are estimated by reweighting the simulated $p_T^{D^*}$ and η^{D^*} distributions to the shape observed in data. The respective uncertainty in the inclusive cross section measurement is 0.5%. This value is taken for the differential measurement as well, due to statistical limitations.

5.3.1.7 Combinatorial Background Normalization

To estimate the uncertainty associated with the rescaling of the wrong charge distribution, two alternative functions are compared to the results of using a combination of a crystal-ball function and a constant function, as described in Sect. 5.2.2.1. A combination of a crystal-ball function with a first order polynomial and a constant function, which was fitted in the off-signal region of $\Delta m(D^0, D^*)$ [0.15, 0.17] are chosen for this test. The uncertainty due to these variations corresponds to 0.5% and is further referred to as 'background normalization'.

5.3.1.8 Missing Transverse Momentum

The uncertainty associated with the determination of $\vec{p}_T^{\text{miss}}$ is estimated on an event-by-event basis [28]. The uncertainty of each event is determined by varying the respective particle-flow objects factoring into the calculation of $\vec{p}_T^{\text{miss}}$ (electrons, photons, muons, jets, etc.) within their associated scale and resolution uncertainties. The uncertainty of $\vec{p}_T^{\text{miss}}$ results in an uncertainty of 0.9% on the inclusive W + c cross section.

5.3.1.9 Pileup

Uncertainties due to the modelling of pileup are estimated by varying the total inelastic cross section used in the simulation of pileup events by 5%. The corresponding uncertainty in the W + c cross section is 2%.

5.3.1.10 Secondary Vertex Fitting

The uncertainty due to the requirement that the tracks associated to a D^0 candidate must originate at a fitted secondary vertex is determined by calculating the D^* reconstruction efficiency in data and MC simulation for events with and without applying this selection criterion. The number of reconstructed D^* candidates after the $OS - SS$ subtraction is compared for events with or without a valid secondary vertex in addition to the proximity requirement ($\Delta_{xy} < 0.1\,\text{cm}$, $\Delta_z < 0.1\,\text{cm}$).

The difference in efficiency between data and MC simulation (ϵ^{Vtx}) is calculated according to Eq. 5.7 and an uncertainty in the inclusive cross section of -1.1% is obtained. Since this variation is not symmetric, the uncertainty is one-sided.

$$\frac{N_{Data}^{D^*}(\text{fitted sec. Vertex})}{N_{Data}^{D^*}(\text{proximity only})} = \epsilon^{Vtx} \cdot \frac{N_{MC}^{D^*}(\text{fitted sec. Vertex})}{N_{MC}^{D^*}(\text{proximity only})} \tag{5.7}$$

5.3.1.11 PDFs

The PDF uncertainties in the cross section measurement are determined, following the prescription of the PDF group. The NNPDF3.0nlo [8] PDF set, used in the signal MC simulation, consists of 100 replicas plus 2 variations of α_S. The cross section of each individual replica is calculated and the standard deviation of the resulting distribution is taken as one PDF uncertainty and corresponds to 1.2%. The uncertainty related to the variation of α_S is 0.08% and is added in quadrature to the PDF uncertainty.

5.3.1.12   Fragmentation

The uncertainty related to the fragmentation modelling in the signal MC is investigated by varying the function, describing the fragmentation parameter $z = p_T^{D^*}/p_T^c$. The procedure employed for this test is based on a measurement of the c $\rightarrow$ D$^*(2010)^{\pm}$ fragmentation function, determined in $e\,p$-collisions, where the resulting parameters were determined with an associated uncertainty of 10% [29].

In Monte Carlo simulations, using the PYTHIA (V8.2.12) event generator, the fragmentation of charm (and bottom) quarks is described by the phenomenological Bowler–Lund function [30, 31], taking on the form:

$$ f(z) = \frac{1}{z^{r_c\,b\,m_q^2}}(1-z)^a \exp(-b\,m_t^2/z)\,c \; , \; \text{with } m_t = \sqrt{m_{D^*}^2 + p_T^{D^*2}} \qquad (5.8) $$

There, the parameter r_c is set to 1 for charm quarks and m_c is the standard setting of the CUETP8M1 underlying event tune, used in the signal sample. The value of m_T corresponds to the average transverse momentum of the generated D$^*(2010)^{\pm}$ mesons. The parameters a and b control the form of the function, whereas c is required for the normalization. All three are determined in a fit to the $p_T^{D^*}/\sum p_T^{\text{track}}$ ratio, with $\sum p_T^{\text{track}}$ corresponding to the sum of track transverse momenta in a cone of $\Delta R \leq 0.4$ around the D* axis. This quantity is used to approximate the charm transverse momentum, as the presence of a jet is not required in this analysis.

The the free parameters a and b are determined in the fit, where the values of $a = 1.827 \pm 0.016$ and $b = 0.00837 \pm 0.00005$ GeV^{-2} are obtained. Both are varied independently by 10% to estimate the systematic uncertainty in the W + c cross section. Figure 5.18 presents the distribution of $p_T^{D^*}/\sum p_T^{\text{track}}$, originating from W + c, the fit of the Bowler–Lund function of the data and the resulting functions for the parameter variations. The variation of the a parameter is restricted to 2, as this is an upper boundary of this parameter, implemented in PYTHIA. However, this restriction is still consistent with a 10% parameter variation.

The resulting shifts in the W + c cross section are added in quadrature and the resulting total fragmentation uncertainty corresponds to 3.9%.

5.3.2   Results

The number of reconstructed D$^*(2010)^{\pm}$ mesons after $OS - SS$ subtraction, the correction factor $\mathcal{C}$ and the inclusive fiducial cross section of W + c with the corresponding uncertainties are listed in Table 5.2. The results are presented for the combination of both W$^{\pm}$ boson charges, as well as the individual cross sections of W$^+ + \bar{c}$ and W$^- + c$, and their associated ratio $\sigma(\text{W}^+ + \bar{c})/\sigma(\text{W}^- + c)$. The correction factors and cross sections, with their associated uncertainties, obtained for the W + D* measurement are presented below the W + c results.

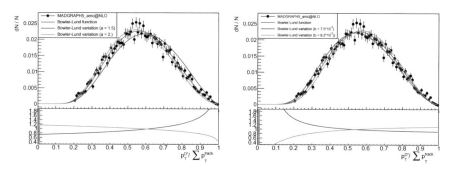

Fig. 5.18 Normalized ratio distribution between $p_T^{D^*}$ and the sum of p_T^{track} in a cone of $\Delta R = 0.4$ around the D* axis for candidates. The Bowler–Lund function [30, 31] is fit to the data points (red line) and variations of the a (left) and b (right) parameters are indicated by blue and green lines

Table 5.2 Fiducial inclusive cross sections of W + c and W + D*(2010)$^\pm$ production in the kinematic range of the analysis. The correction factor $\mathcal{C}$ accounts for the acceptance and efficiency of the detector

	W + c	W$^+$ + c̄	W$^-$ + c
N_{sel}	19210 ± 587 stat.	9674 ± 401 stat.	9546 ± 367 stat.
$\mathcal{C}$	0.0811 ± 0.003 stat.	0.0832 ± 0.004 stat.	0.0794 ± 0.004 stat.
σ (pb)	1026 ± 31 stat. $^{+76}_{-72}$ syst.	504 ± 21 stat. ± 42 syst.	521 ± 20 stat. $^{+42}_{-40}$ syst.
$\frac{\sigma(W^+ + \bar{c})}{\sigma(W^- + c)}$		0.968 ± 0.055 stat. $^{+0.015}_{-0.028}$	
	W + D*	W$^+$ + D*(2010)$^-$	W$^-$ + D*(2010)$^+$
N_{sel}	19210 ± 587 stat.	9674 ± 401 stat.	9546 ± 367 stat.
$\mathcal{C}$	0.107 ± 0.004 stat.	0.113 ± 0.006 stat.	0.101 ± 0.004 stat.
σ (pb)	190 ± 6 stat. $^{+12}_{-13}$ syst.	90 ± 4 stat. $^{+7}_{-8}$ syst.	99 ± 3 stat. ± 7 syst.
$\frac{\sigma(W^+ + D^*(2010)^-)}{\sigma(W^- + D^*(2010)^+)}$		0.909 ± 0.051 stat. $^{+0.014}_{-0.028}$	

The measured values of the differential W + c measurement are summarized in Table 5.3. The number of signal events after $OS - SS$ subtraction are listed together with the correction factor $\mathcal{C}$ in each bin, derived from MC simulation, and the cross section with the corresponding uncertainties. The results are presented for $d\sigma(W + c)/d|\eta^\mu|$, as well as for $d\sigma(W^+ + \bar{c})/d|\eta^\mu|$ and $d\sigma(W^- + c)/d|\eta^\mu|$. The values of the differential W + D* cross sections are listed in Table 5.4, together with the corrections factors $\mathcal{C}$ and associated uncertainties.

Table 5.3 Number of reconstructed $D^*(2010)^{\pm}$ mesons after $OS - SS$, correction factors C, accounting for the acceptance and efficiency of the detector, and the fiducial cross sections in each range of $|\eta^{\mu}|$ for $W + c$ (upper), $W^+ + \bar{c}$ (middle) and $W^- + c$ (lower)

W+c

$[|\eta^{\mu}_{\min}|, |\eta^{\mu}_{\max}|]$	N_{sel}	C	$\frac{d\sigma(W+c)}{d|\eta^{\mu}|}$ (pb)
$[0, 0.4]$	$3795 \pm 248\,\text{stat.}$	$0.072 \pm 0.006\,\text{stat.}$	$569 \pm 37\,\text{stat.}\,^{+61}_{-53}$
$[0.4, 0.8]$	$4201 \pm 256\,\text{stat.}$	$0.096 \pm 0.006\,\text{stat.}$	$467 \pm 28\,\text{stat.}\,^{+61}_{-66}$
$[0.8, 1.3]$	$4334 \pm 274\,\text{stat.}$	$0.078 \pm 0.006\,\text{stat.}$	$479 \pm 30\,\text{stat.}\,^{+49}_{45}$
$[1.3, 1.8]$	$3823 \pm 267\,\text{stat.}$	$0.083 \pm 0.007\,\text{stat.}$	$395 \pm 28\,\text{stat.}\,^{+49}_{-63}$
$[1.8, 2.4]$	$3042 \pm 266\,\text{stat.}$	$0.078 \pm 0.007\,\text{stat.}$	$283 \pm 25\,\text{stat.}\,^{+39}_{-34}$

$W^+ + \bar{c}$

$[|\eta^{\mu}_{\min}|, |\eta^{\mu}_{\max}|]$	N_{sel}	C	$\frac{d\sigma(W^+ + \bar{c})}{d|\eta^{\mu}|}$ (pb)
$[0, 0.4]$	$2109 \pm 167\,\text{stat.}$	$0.073 \pm 0.008\,\text{stat.}$	$313 \pm 25\,\text{stat.}\,^{+48}_{-44}$
$[0.4, 0.8]$	$2119 \pm 172\,\text{stat.}$	$0.094 \pm 0.010\,\text{stat.}$	$236 \pm 19\,\text{stat.}\,^{+37}_{-41}$
$[0.8, 1.3]$	$2103 \pm 186\,\text{stat.}$	$0.077 \pm 0.008\,\text{stat.}$	$235 \pm 21\,\text{stat.}\,^{+33}_{-27}$
$[1.3, 1.8]$	$1840 \pm 184\,\text{stat.}$	$0.093 \pm 0.010\,\text{stat.}$	$162 \pm 16\,\text{stat.}\,^{+34}_{-31}$
$[1.8, 2.4]$	$1499 \pm 186\,\text{stat.}$	$0.080 \pm 0.011\,\text{stat.}$	$135 \pm 17\,\text{stat.}\,^{+24}_{-26}$

$W^- + c$

$[|\eta^{\mu}_{\min}|, |\eta^{\mu}_{\max}|]$	N_{sel}	C	$\frac{d\sigma(W^- + c)}{d|\eta^{\mu}|}$ (pb)
$[0, 0.4]$	$1688 \pm 158\,\text{stat.}$	$0.072 \pm 0.008\,\text{stat.}$	$255 \pm 23\,\text{stat.}\,^{+35}_{-42}$
$[0.4, 0.8]$	$2084 \pm 162\,\text{stat.}$	$0.097 \pm 0.008\,\text{stat.}$	$231 \pm 18\,\text{stat.}\,^{+28}_{-42}$
$[0.8, 1.3]$	$2234 \pm 172\,\text{stat.}$	$0.079 \pm 0.007\,\text{stat.}$	$244 \pm 19\,\text{stat.}\,^{+29}_{-38}$
$[1.3, 1.8]$	$1986 \pm 166\,\text{stat.}$	$0.073 \pm 0.008\,\text{stat.}$	$237 \pm 20\,\text{stat.}\,^{+33}_{-37}$
$[1.8, 2.4]$	$1544 \pm 161\,\text{stat.}$	$0.075 \pm 0.008\,\text{stat.}$	$149 \pm 16\,\text{stat.}\,^{+25}_{-21}$

5.3.3 Comparisons with Theoretical Predictions

The results of the inclusive and differential $W + c$ measurement are compared to predictions calculated with MCFM 6.8 [1–3] at NLO ($\mathcal{O}(\alpha_s^2)$), in combination with different PDF sets. The mass of the c quark is set to $m_c = 1.5\,\text{GeV}$ and refers to the pole mass. The value corresponds to the charm mass setting in the signal MC. The calculations are performed for $p_T^{\mu} > 26\,\text{GeV}$, $\eta^{-} < 2.4$, and $p_T^c > 5\,\text{GeV}$, corresponding to the kinematic requirements of the analysis at the generator level. The strong coupling constant $\alpha_S(m_Z)$ is set to the value used in the evaluation of the particular PDF, and the factorization and renormalization scales are set to $\mu_f = \mu_r = m_W$. The PDF uncertainties are determined using the prescriptions provided by each PDF group, including the variations of α_S, if present. The prediction values for the inclusive cross sections and their respective PDF uncertainties are listed in Table 5.5, with $\Delta\mu$ referring to scale uncertainties, conventionally used to estimate the effect of higher order corrections. These are evaluated by simultaneously varying μ_f and μ_r by a factor of 2 up and down.

Table 5.4 Number of reconstructed $D^*(2010)^\pm$ mesons after $OS - SS$, correction factors $\mathcal{C}$, accounting for the acceptance and efficiency of the detector and the fiducial cross sections in each range of $|\eta^\mu|$ for $W + D^*(2010)^\pm$ (upper), $W^+ + D^*(2010)^-$ (middle) and $W^- + D^*(2010)^+$ (lower)

$W + D^*(2010)^\pm$

$[|\eta^\mu_{min}|, |\eta^\mu_{max}|]$	N_{sel}	$\mathcal{C}$	$\frac{d\sigma(W+D^*(2010)^\pm)}{d|\eta^\mu|}$ (pb)
$[0, 0.4]$	3795 ± 248	0.137 ± 0.011	$72.9 \pm 4.8\,\text{stat.}\,^{+7.7}_{-6.6}$
$[0.4, 0.8]$	4201 ± 256	0.176 ± 0.012	$62.2 \pm 3.8\,\text{stat.}\,^{+11.5}_{-12.5}$
$[0.8, 1.3]$	4334 ± 274	0.141 ± 0.011	$64.7 \pm 4.1\,\text{stat.}\,^{+6.4}_{-5.8}$
$[1.3, 1.8]$	3823 ± 267	0.154 ± 0.012	$51.6 \pm 3.6\,\text{stat.}\,^{+5.9}_{-8.3}$
$[1.8, 2.4]$	3042 ± 266	0.139 ± 0.013	$38.3 \pm 3.4\,\text{stat.}\,^{+7.2}_{-6.7}$

$W^+ + D^*(2010)^-$

$[|\eta^\mu_{min}|, |\eta^\mu_{max}|]$	N_{sel}	$\mathcal{C}$	$\frac{d\sigma(W^++D^*(2010)^-)}{d|\eta^\mu|}$ (pb)
$[0, 0.4]$	2109 ± 167	0.140 ± 0.016	$39.6 \pm 3.1\,\text{stat.}\,^{+6.1}_{-5.5}$
$[0.4, 0.8]$	2119 ± 172	0.174 ± 0.017	$32.0 \pm 2.6\,\text{stat.}\,^{+3.8}_{-5.1}$
$[0.8, 1.3]$	2103 ± 186	0.141 ± 0.015	$31.5 \pm 2.8\,\text{stat.}\,^{+4.3}_{-3.6}$
$[1.3, 1.8]$	1840 ± 184	0.176 ± 0.019	$19.4 \pm 2.0\,\text{stat.}\,^{+5.1}_{-3.8}$
$[1.8, 2.4]$	1499 ± 186	0.151 ± 0.021	$17.3 \pm 2.2\,\text{stat.}\,^{+3.1}_{-3.8}$

$W^- + D^*(2010)^+$

$[|\eta^\mu_{min}|, |\eta^\mu_{max}|]$	N_{sel}	$\mathcal{C}$	$\frac{d\sigma(W^-+D^*(2010)^+)}{d|\eta^\mu|}$ (pb)
$[0, 0.4]$	1688 ± 158	0.135 ± 0.015	$33.1 \pm 3.1\,\text{stat.}\,^{+4.4}_{-5.5}$
$[0.4, 0.8]$	2084 ± 162	0.175 ± 0.015	$31.3 \pm 2.4\,\text{stat.}\,^{+3.7}_{-4.5}$
$[0.8, 1.3]$	2234 ± 172	0.142 ± 0.013	$33.2 \pm 2.5\,\text{stat.}\,^{+3.9}_{-3.8}$
$[1.3, 1.8]$	1986 ± 166	0.132 ± 0.015	$31.4 \pm 2.6\,\text{stat.}\,^{+4.3}_{-4.5}$
$[1.8, 2.4]$	1544 ± 161	0.129 ± 0.014	$21.0 \pm 2.2\,\text{stat.}\,^{+3.5}_{-2.9}$

Table 5.5 The fixed order predictions for $\sigma(W + c)$, obtained with MCFM [1–3] and using different PDFs. The uncertainties account for PDF and scale variations

	$\sigma(W + c)$ (pb)	ΔPDF (%)	$\Delta\mu$ (%)	$\sigma(W^+ + \bar{c})/\sigma(W^- + c)$
ABMP16nlo	1077.9	± 2.1	$^{+3.4}_{-2.4}$	$0.975\,^{+0.002}_{-0.002}$
ATLASepWZ16nnlo	1235.1	$^{+1.4}_{-1.6}$	$^{+3.7}_{-2.8}$	$0.976\,^{+0.001}_{-0.001}$
CT14nlo	992.6	$^{+7.2}_{-8.4}$	$^{+3.1}_{-2.1}$	$0.970\,^{+0.005}_{-0.007}$
MMHT14nlo	1057.1	$^{+6.5}_{-8.0}$	$^{+3.2}_{-2.2}$	$0.960\,^{+0.023}_{-0.033}$
NNPDF3.0nlo	959.5	± 5.4	$^{+2.8}_{-1.9}$	$0.962\,^{+0.034}_{-0.034}$
NNPDF3.1nlo	1030.2	± 5.3	$^{+3.2}_{-2.2}$	$0.965\,^{+0.043}_{-0.043}$

Figure 5.19 presents the measurements of the inclusive W + c, W$^+$ + c̄, and W$^-$ + c cross sections, as well as the charge ratio, comp ared to NLO predictions using the ABMP16nlo [32], ATLASepWZ16nnlo [33], CT14nlo [34], MMHT14nlo [35], NNPDF3.0nlo [8], and NNPDF3.1nlo [36] PDF sets. The prediction obtained with ATLASepWZ16nnlo is the only one using a NNLO PDF set, as no corresponding NLO dataset is currently available. Furthermore, only Hessian uncertainties are considered for a direct comparison with the other datasets. By con-

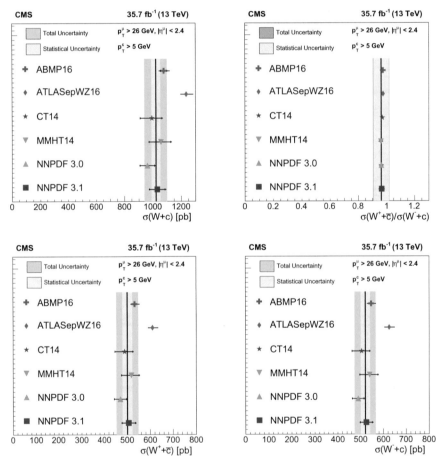

Fig. 5.19 Inclusive fiducial cross section of W + c at 13 TeV (upper left), cross section ratio $\sigma(W^+ + \bar{c})/\sigma(W^- + c)$ (upper right), $\sigma(W^+ + \bar{c})$ (lower left) and W$^-$ + c (lower right). The cross section (ratio) values are represented by a black line, while the statistical and total uncertainties of the measurement are shown as shaded bands. The measurements are compared to fixed order QCD predictions, done with MCFM at NLO, using various PDF sets, represented by different types of symbols. The PDFs are evaluated at NLO, with the exception of ATLASepWZ16nnlo, which has been determined at NNLO. The error bars represent the total theoretical uncertainty, containing the uncertainties related to the PDF and the scale variations

vention, the PDF uncertainties provided by the CTEQ collaboration correspond to 90% confidence level (CL), whereas the uncertainties of the other PDF sets used for the theoretical predictions are given at 68% CL. In order to have a uniform representation of the PDF uncertainties, those obtained with CT14nlo are scaled down accordingly. The datasets used in the determination of the strangeness distribution in each of the used PDF sets are discussed in Sect. 2.2.5. However, it is worthwhile to mention once again that the ABMP16nlo set includes the latest results of the NOMAD and CHORUS experiments in their fit. These measurements probe the strange quark content of the nucleon via charm production in charged-current neutrino-nucleon DIS, resulting a small uncertainty associated with the strange quark distribution. The comparison with NNPDF3.0nlo is presented in addition to NNPDF3.1nlo as it is the PDF set used in the signal simulation of the analysis. The results obtained by using either PDF set in a prediction differ from each other due to the additional measurements included in the fit of NNPDF3.1nlo. Especially the inclusion of the inclusive W and Z^0 measurements at 7 TeV performed by ATLAS at full statistics causes a shift in the strange quark distribution, as is demonstrated in Fig. 4.19 of reference [36]. Despite the differences in methodology, the strangeness suppression distributions of the ABMP16nlo, CT14nlo, MMHT14nlo and NNPDF3.1nlo PDF sets agree with each other and disagree with the results of ATLASepWZ16nnlo. Good agreement between the inclusive W + c, $W^+ + \bar{c}$ and $W^- + c$ measurements and the fixed order predictions is observed, except for the calculation using the ATLASepWZ16nnlo PDF set. Tables 5.6 and 5.7 summarize the prediction values, using different PDFs for the differential measurement $d\sigma(W + c)/d|\eta^\mu|$. The small asymmetry between $\sigma(W^+ + \bar{c})$ and $\sigma(W^- + c)$ is related to the contributions of d valence quark to the $W^- + c$ cross section and good agreement between the measurement and all theoretical predictions is observed.

The W + D* cross section is compared to the signal MC prediction, which uses NNPDF3.0nlo to describe the proton structure. The PDF and α_S uncertainties are accounted for and the scale uncertainties are determined by varying μ_r and μ_f in the matrix element by a factor of 2. The prediction values for the differential W + D* cross section and their respective PDF uncertainties are listed in Table 5.8. Figure 5.20 presents the differential W + c and W + D* cross sections, compared with the fixed order calculations done with MCFM and the signal MC prediction, respectively. The measured cross section is in good agreement with the simulation.

Table 5.6 Theoretical predictions for $d\sigma(W + c)/\ d|\eta^\mu|$ done with MCFM 6.8, using the ABMP16nlo, ATLASepWZ16nnlo and CT14nlo PDF sets. The associated PDF and scale uncertainties are shown relative to the prediction values

ABMP16nlo

$[|\eta^\mu_{min}|, |\eta^\mu_{max}|]$	$\frac{d\sigma(W+c)}{d|\eta^\mu|}$ (pb)	ΔPDF (%)	$\Delta\mu$ (%)
[0, 0.4]	537.8	± 2.2	$^{+3.7}_{-1.9}$
[0.4, 0.8]	522.8	± 2.1	$^{+3.1}_{-2.3}$
[0.8, 1.3]	483.9	± 2.1	$^{+3.2}_{-2.1}$
[1.3, 1.8]	422.4	± 2.0	$^{+3.4}_{-2.9}$
[1.8, 2.4]	334.1	± 2.0	$^{+3.4}_{-3.0}$

ATLASepWZ16nnlo

$[|\eta^\mu_{min}|, |\eta^\mu_{max}|]$	$\frac{d\sigma(W+c)}{d|\eta^\mu|}$ (pb)	ΔPDF (%)	$\Delta\mu$ (%)
[0, 0.4]	607.8	$^{+1.1}_{-1.3}$	$^{+4.2}_{-2.4}$
[0.4, 0.8]	592.9	$^{+1.1}_{-1.3}$	$^{+3.5}_{-2.7}$
[0.8, 1.3]	552.7	$^{+1.2}_{-1.4}$	$^{+3.6}_{-2.5}$
[1.3, 1.8]	487.8	$^{+1.4}_{-1.6}$	$^{+3.8}_{-3.3}$
[1.8, 2.4]	391.1	$^{+2.2}_{-2.3}$	$^{+3.6}_{-3.3}$

CT14nlo

$[|\eta^\mu_{min}|, |\eta^\mu_{max}|]$	$\frac{d\sigma(W+c)}{d|\eta^\mu|}$ (pb)	ΔPDF (%)	$\Delta\mu$ (%)
[0, 0.4]	499.3	$^{+7.0}_{-8.0}$	$^{+3.4}_{-1.7}$
[0.4, 0.8]	484.4	$^{+7.0}_{-8.0}$	$^{+2.9}_{-2.1}$
[0.8, 1.3]	446.3	$^{+6.9}_{-8.2}$	$^{+2.9}_{-1.8}$
[1.3, 1.8]	387.0	$^{+7.1}_{-8.5}$	$^{+3.1}_{-2.6}$
[1.8, 2.4]	304.1	$^{+7.8}_{-9.3}$	$^{+3.0}_{-2.6}$

Table 5.7 Theoretical predictions for $d\sigma(W+c)/d|\eta^{\mu}|$ done with MCFM 6.8, using the MMHT14nlo, NNPDF3.0nlo and NNPDF3.1nlo PDF sets. The associated PDF and scale uncertainties are shown relative to the prediction values

MMHT14nlo

$[|\eta^{\mu}_{min}|, |\eta^{\mu}_{max}|]$	$\frac{d\sigma(W+c)}{d|\eta^{\mu}|}$ (pb)	ΔPDF (%)	$\Delta\mu$ (%)
[0, 0.4]	526.0	$+7.0$ -7.7	$+3.6$ -1.8
[0.4, 0.8]	511.2	$+6.8$ -7.7	$+3.0$ -2.1
[0.8, 1.3]	473.4	$+6.4$ -7.7	$+3.0$ -1.9
[1.3, 1.8]	414.4	$+6.0$ -8.0	$+3.2$ -2.7
[1.8, 2.4]	330.5	$+6.5$ -9.1	$+3.2$ -2.7

NNPDF3.0nlo

$[|\eta^{\mu}_{min}|, |\eta^{\mu}_{max}|]$	$\frac{d\sigma(W+c)}{d|\eta^{\mu}|}$ (pb)	ΔPDF (%)	$\Delta\mu$ (%)
[0, 0.4]	489.8	± 7.0	$+3.2$ -1.5
[0.4, 0.8]	473.2	± 6.5	$+2.7$ -1.8
[0.8, 1.3]	432.4	± 5.5	$+2.6$ -1.5
[1.3, 1.8]	370.4	± 4.2	$+2.7$ -2.3
[1.8, 2.4]	288.1	± 3.5	$+2.7$ -2.3

NNPDF3.1nlo

$[|\eta^{\mu}_{min}|, |\eta^{\mu}_{max}|]$	$\frac{d\sigma(W+c)}{d|\eta^{\mu}|}$ (pb)	ΔPDF (%)	$\Delta\mu$ (%)
[0, 0.4]	524.8	± 5.8	$+3.6$ -1.8
[0.4, 0.8]	508.1	± 5.6	$+3.0$ -2.2
[0.8, 1.3]	465.6	± 5.4	$+3.0$ -1.9
[1.3, 1.8]	399.0	± 5.0	$+3.1$ -2.7
[1.8, 2.4]	307.9	± 4.8	$+3.1$ -2.6

Table 5.8 Theoretical predictions for $d\sigma(W+D^{*\pm})/d|\eta^{\mu}|$ calculated with MADGRAPH5_aMC@NLO. The relative uncertainty due to PDF and scale variations are indicated

$[|\eta_{min}|, |\eta_{max}|]$	MADGRAPH5_aMC@NLO + PYTHIA		
	$\frac{d\sigma(W+D^*(2010)^{\pm})}{d|\eta^{\mu}|}$ (pb)	ΔPDF (%)	$\Delta\mu$ (%)
[0, 0.4]	72.9	± 6.4	$+0.4$ -0.2
[0.4, 0.8]	70.2	± 5.9	$+0.4$ -0.4
[0.8, 1.3]	64.0	± 5.0	$+0.5$ -0.4
[1.3, 1.8]	55.6	± 3.8	$+0.5$ -0.1
[1.8, 2.4]	43.6	± 3.6	$+1.2$ -0.7

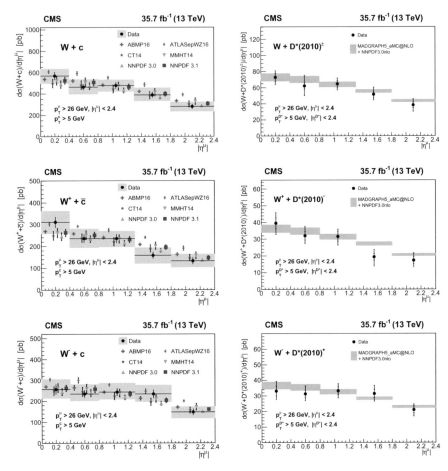

Fig. 5.20 Left: Differential cross section of $\sigma(\mathrm{W}^+ + \mathrm{c})$ (upper) $\sigma(\mathrm{W}^+ + \bar{\mathrm{c}})$ (middle) and $\sigma(\mathrm{W}^- + \mathrm{c})$ (lower) as a function of $|\eta^\mu|$. The measured cross sections are represented by black points with the vertical bars indicating the statistical uncertainties, whereas the coloured boxes correspond to the total uncertainty. The results are compared to theoretical predictions done with MCFM [1–3] at NLO using different PDF sets, indicated by symbols of different forms and colours. Right: Differential cross section of $\mathrm{W} + \mathrm{D}^*(2010)^\pm$ as a function of $|\eta^\mu|$. The results of the measurement are indicated by black points, with the inner error bars representing the statistical uncertainties and the outer bars corresponding to the total uncertainty. The data is compared to a prediction done using the signal MC sample, generated with MADGRAPH5_aMC@NLO, and the proton structure is described by the NNPDF3.0nlo PDF set

References

1. Campbell JM, Ellis RK (2010) MCFM for the Tevatron and the LHC. Nucl Phys Proc Suppl 205–206:10. https://doi.org/10.1016/j.nuclphysbps.2010.08.011. arXiv:1007.3492 [hep-ph]
2. Campbell JM, Keith Ellis R (1999) An update on vector boson pair production at hadron colliders. Phys Rev D 60:113006. https://doi.org/10.1103/PhysRevD.60.113006. arXiv:hep-ph/9905386 [hep-ph]

3. Campbell JM, Keith Ellis R (2015) Top quark processes at NLO in production and decay. J Phys G 42:015005. https://doi.org/10.1088/0954-3899/42/1/015005. arXiv:1204.1513 [hep-ph]
4. Buckley A et al (2013) Rivet user manual. Comput Phys Commun 184:2803. https://doi.org/10.1016/j.cpc.2013.05.021. arXiv:1003.0694 [hep-ph]
5. Rovere M (2015) The data quality monitoring software for the CMS experiment at the LHC. J Phys Conf Ser 664(7):072039. http://stacks.iop.org/1742-6596/664/i=7/a=072039
6. The CMS Collaboration. CMS Luminosity Measurements for the 2016 Data Taking Period. CMS Physics Analysis Summary CMS-PAS-LUM-17-001. 2017. http://cds.cern.ch/record/2257069
7. Agostinelli et al (2003) Geant4 - a simulation toolkit. Nucl Instr Meth A 506:250. ISSN: 0168-9002. https://doi.org/10.1016/S0168-9002(03)01368-8. https://www.sciencedirect.com/science/article/pii/S0168900203013688
8. Ball RD et al (2015) Parton distributions for the LHC Run II. JHEP 04:040. https://doi.org/10.1007/JHEP04(2015)040. arXiv:1410.8849 [hep-ph]
9. Sjöstrand T, Mrenna S, Skands PZ (2008) A brief introduction to PYTHIA 8.1. Comput Phys Commun 178:852. https://doi.org/10.1016/j.cpc.2008.01.036. arXiv:0710.3820 [hep-ph]
10. Khachatryan V et al (2016) Event generator tunes obtained from underlying event and multiparton scattering measurements. Eur Phys J C 76:155. https://doi.org/10.1140/epjc/s10052-016-3988-x. arXiv:1512.00815 [hep-ex]
11. Alwall J et al (2014) The automated computation of tree-level and next-to-leading order differential cross sections, and their matching to parton shower simulations. JHEP 07:079. https://doi.org/10.1007/JHEP07(2014)079. arXiv:1405.0301 [hep-ph]
12. Frederix R, Frixione S (2012) Merging meets matching in MC@NLO. JHEP 12:061. https://doi.org/10.1007/JHEP12(2012)061. arXiv:1209.6215 [hep-ph]
13. Campbell JM et al (2015) Top-pair production and decay at NLO matched with parton showers. JHEP 04:114. https://doi.org/10.1007/JHEP04(2015)114. arXiv:1412.1828 [hep-ph]
14. Investigations of the impact of the parton shower tuning in Pythia 8 in the modelling of tt at ps = 8 and 13 TeV. CMS Physics Analysis Summary (2016). https://cds.cern.ch/record/2235192
15. Re E (2011) Single-topWt-channel production matched with parton showers using the POWHEG method. Eur Phys J C 71:1547. https://doi.org/10.1140/epjc/s10052-011-1547-z. arXiv:1009.2450 [hep-ph]
16. Frederix R, Re E, Torrielli P (2012) Single-top t-channel hadroproduction in the four-flavour scheme with POWHEG and aMC@NLO. JHEP 09:130. https://doi.org/10.1007/JHEP09(2012)130. arXiv:1207.5391 [hep-ph]
17. Alioli S, et al (2009) NLO single-top production matched with shower in POWHEG: s- and t-channel contributions. JHEP 09:111. [Erratum: (2010) JHEP 02:011]. https://doi.org/10.1007/JHEP02(2010)011, https://doi.org/10.1088/1126-6708/2009/09/111. arXiv:0907.4076 [hep-ph]
18. Particle Data Group, Tanabashi M, et al (2018) Review of particle physics. Phys Rev D 98:030001. https://doi.org/10.1103/PhysRevD.98.030001
19. Sirunyan AM, et al (2017) Particle-flow reconstruction and global event description with the CMS detector. JINST 12:P10003. https://doi.org/10.1088/1748-0221/12/10/P10003. arXiv:1706.04965 [physics.ins-det]
20. Sirunyan AM, et al (2018) Performance of the CMS muon detector and muon reconstruction with proton-proton collisions at ps = 13 TeV. JINST 13:P06015. https://doi.org/10.1088/1748-0221/13/06/P06015. arXiv:1804.04528 [physics.ins-det]
21. Frühwirth R, Waltenberger W, Vanlaer P (2007) Adaptive vertex fitting. Technical report CMS-NOTE-2007-008. https://cds.cern.ch/record/1027031
22. Chatrchyan S, et al (2014) Description and performance of track and primary vertex reconstruction with the CMS tracker. JINST 9:P10009. https://doi.org/10.1088/1748-0221/9/10/P10009. arXiv:1405.6569 [physics.ins-det]
23. Feldman GJ et al (1977) Observation of the decay $D^{*}+ \rightarrow D^{0}\pi^{+}$. Phys Rev Lett 38:1313. https://doi.org/10.1103/PhysRevLett.38.1313

24. Oreglia MJ (1980) A study of the reactions $\psi' \to \gamma\gamma\psi$! ". SLAC Report SLAC-R-236. PhD thesis. Stanford University. http://www.slac.stanford.edu/cgi-wrap/getdoc/slac-r-236.pdf

25. Lisovyi M, Verbytskyi A, Zenaiev O (2016) Combined analysis of charm-quark fragmentation-fraction measurements. Eur Phys J C 76:397. https://doi.org/10.1140/epjc/s10052-016-4246-y. arXiv:1509.01061 [hep-ex]

26. The CMS Collaboration. Measurement of tracking efficiency. CMS Physics Analysis Summary CMS-PAS-TRK-10-002 (2010). http://cds.cern.ch/record/1279139

27. The CMS Collaboration. CMS Tracking POG Performance Plots for year 2016. https://twiki.cern.ch/twiki/bin/view/CMSPublic/TrackingPOGPlots2016

28. The CMS Collaboration. Performance of missing energy reconstruction in 13 TeV pp collision data using the CMS detector. CMS Physics Analysis Summary CMS-PAS-JME-16-004 (2016). https://cds.cern.ch/record/2205284

29. Aaron FD et al (2009) Study of charm fragmentation into $D_\pm$ mesons in deepinelastic scattering at HERA. Eur Phys J C 59:589. https://doi.org/10.1140/epjc/s10052-008-0792-2. arXiv:0808.1003 [hep-ex]

30. Andersson B et al (1983) Parton fragmentation and string dynamics. Phys Rept 97:31. https://doi.org/10.1016/0370-1573(83)90080-7

31. Bowler MG (1981) e^+e^- Production of heavy quarks in the string model. Z Phys C 11:169. https://doi.org/10.1007/BF01574001

32. Alekhin S, Blümlein J, Moch S (2018) NLO PDFs from the ABMP16 fit. Eur Phys J C 78:477. https://doi.org/10.1140/epjc/s10052-018-5947-1. arXiv:1803.07537 [hep-ph]

33. Aaboud RD et al (2017) Precision measurement and interpretation of inclusive W^+, W^- and $Z/\gamma*$ production cross sections with the ATLAS detector. Eur Phys J C 77:367. https://doi.org/10.1140/epjc/s10052-017-4911-9. arXiv:1612.03016 [hep-ex]

34. Dulat S et al (2016) New parton distribution functions from a global analysis of quantum chromodynamics. Phys Rev D 93:033006. https://doi.org/10.1103/PhysRevD.93.033006. arXiv:1506.07443 [hep-ph]

35. Harland-Lang LA et al (2015) Parton distributions in the LHC era: MMHT 2014 PDFs. Eur Phys J C 75:204. https://doi.org/10.1140/epjc/s10052-015-3397-6. arXiv:412.3989 [hep-ph]

36. Ball RD et al (2017) Parton distributions from high-precision collider data. Eur Phys J C 77:663. https://doi.org/10.1140/epjc/s10052-017-5199-5. arXiv:1706.00428 [hep-ph]

Chapter 6
Determination of the Strange Quark Content of the Proton

The results for the associated production of a $W^{\pm}$ boson and a charm quark at 13 TeV, presented in the previous chapter, are included in a QCD analysis to illustrate the impact of the measurement on possible constraints of the strange quark content of the proton. The QCD analysis presented in this chapter is performed to NLO accuracy, utilizing the XFITTER [1, 2] framework, version 2.0.0. The most recent HERA DIS data in combination with the CMS measurements on the $W^{\pm}$ boson asymmetry and $W + c$ are used. The input datasets, chosen parametrization and model assumptions are presented, and the results are compared with the PDFs obtained in earlier analyses. Furthermore, for the sake of comparison to the earlier CMS result [3], the fit is repeated using the same parametrization as in [3].

6.1 Datasets Used in the PDF Fit

The most recent combination of the HERA I+II measurements [4] of CC and NC processes in $e^{\pm}p$ collisions are included in the fit due to their sensitivity to light quarks in the low and medium regions of Bjorken x, and to the gluon via scaling violations. The dataset is a combination of all inclusive DIS cross sections measured by the H1 and ZEUS collaborations during the full runtime of the HERA accelerator and includes results from different centre-of-mass energies. Extensive evaluations of the correlated and uncorrelated systematic uncertainties were performed to ensure the consistency of the results [4]. The NC cross sections have been determined in the kinematic range of $6 \cdot 10^{-7} \le x \le 0.65$, whereas the CC cross sections are available for the kinematic range of $1.3 \cdot 10^{-2} \le x \le 0.40$. While both, NC and CC, are sensitive to the valence quark distributions at high values of Q^2, the measurements of NC cross sections in the low Q^2 region also constrain information on the sea quark

© The Editor(s) (if applicable) and The Author(s), under exclusive license
to Springer Nature Switzerland AG 2020
S. K. Pflitsch, *Associated Production of W + Charm in 13 TeV Proton-Proton Collisions Measured with CMS and Determination of the Strange Quark Content of the Proton*, Springer Theses,
https://doi.org/10.1007/978-3-030-52762-4_6

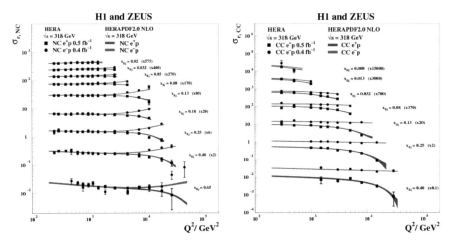

Fig. 6.1 NC (left) and CC (right) cross sections as a function of Q^2, evaluated for different values of Bjorken-x, determined from the combined HERA data in e^+p and e^-p collisions at $\sqrt{s} = 318$ GeV and compared to predictions based on the HERAPDF 2.0 PDF set [4]

distribution in the low x region. The gluon distribution is probed via the scaling violations. Figure 6.1 presents the NC and CC cross sections determined from the combined HERA data in e^+p and e^-p collisions at $\sqrt{s} = 318$ GeV.

Along with the HERA DIS data, both currently available CMS measurements on the lepton charge asymmetry in W$^\pm$ boson production for $\sqrt{s} = 7$ TeV [3] and $\sqrt{s} = 8$ TeV [5] are included due to their sensitivity to the valence quark distribution in the kinematic region of $10^{-3} \leq x \leq 10^{-1}$. The dominant production process for W$^\pm$ bosons in pp-collisions is the hard scattering of a valence quark with a different flavour sea anti-quark which carries the same charge e.g. (u$_v$ + $\bar{\text{d}}$ → W$^+$, d$_v$ + $\bar{\text{u}}$ → W$^-$). Due to the valence quark composition of the proton, the production of W$^+$ bosons has a higher cross section. This can be observed in the distribution of the lepton charge asymmetry, which is defined as follows for leptonic W$^\pm$ decays:

$$A(\eta) = \frac{\frac{d\sigma}{d\eta}(W^+ \to l^+ + \nu) - \frac{d\sigma}{d\eta}(W^- \to l^- + \bar{\nu})}{\frac{d\sigma}{d\eta}(W^+ \to l^+ + \nu) + \frac{d\sigma}{d\eta}(W^- \to l^- + \bar{\nu})} \tag{6.1}$$

Moreover, the W$^\pm$ boson cross sections provide an indirect sensitivity to the strange quark distribution through processes like u$_v$ + $\bar{\text{s}}$ → W$^+$ and $\bar{\text{u}}$ + s → W$^-$. Figure 6.2 presents the lepton charge asymmetry in W$^\pm$ production as a function of the lepton pseudorapidity measured by CMS for centre-of-mass energies of $\sqrt{s} = 7$ TeV and $\sqrt{s} = 8$ TeV compared to predictions, calculated using FEWZ 3.1 [6] in combination with different PDF sets at NLO (7 TeV) or NNLO (8 TeV).

In addition to the differential CMS W + c measurement at $\sqrt{s} = 13$ TeV, presented in Chap. 5, the earlier measurement at $\sqrt{s} = 7$ TeV [7] is included in the QCD fit. Both analyses measure the differential cross section of W + c as a func-

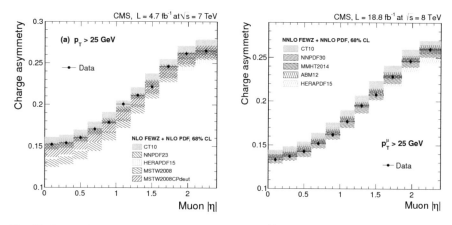

Fig. 6.2 Lepton charge asymmetry, measured by CMS at $\sqrt{s} = 7$ TeV [3] (left) and $\sqrt{s} = 8$ TeV [5] (right) compared to predictions, calculated using FEWZ [6] in combination with different PDF sets at NLO (7 TeV) or NNLO (8 TeV)

tion of the absolute pseudorapidity of the lepton $|\eta^l|$ originating from the $W^{\pm}$ boson decay.

The theoretical calculations for the CMS measurements were done using MCFM [8–10] at NLO ($\mathcal{O}(\alpha_s^2)$) interfaced with APPLGRID 1.4.56 [11]. The predictions of the $W^{\pm}$ asymmetry measurements and the 7 TeV W + c analysis are publicly available via HEPForge [12], while the calculations of for the 13 TeV W + c measurement were done in the course of the analysis, presented in this thesis. The correlated systematic uncertainties of each individual measurement are taken into account, though the individual datasets are treated as uncorrelated to each other. The CMS measurements are treated as uncorrelated due to differences in the event reconstruction and selection, used in the analysis at different centre of mass energies.

The parton distributions are evolved from a starting scale μ_0, using the NLO DGLAP equations, as they are implemented in the QCDNUM 17-00/06 [13] program, which performs all calculations in the $\overline{MS}$ scheme [14, 15]. The heavy quark masses are treated according to the NLO Thorne–Roberts [16, 17] general-mass variable flavour scheme, with $m_b = 4.5$ GeV and $m_c = 1.5$ GeV, corresponding to the settings of the simulated event sample of the W + c measurement. The value of the strong coupling constant is set to $\alpha_S = 0.118$ and the renormalization and factorization scales are defined as $\mu_r = \mu_f = Q$. In the case of DIS data Q corresponds to the four-momentum transfer between the lepton and the parton, while it is set to the mass of the $W^{\pm}$ boson for the $W^{\pm}$ asymmetry data and the W + c process. Following the strategy of the HERAPDF 2.0 QCD analysis, the Q^2 range of the HERA DIS data is restricted to $Q^2 \geq 3.5$ GeV^2 to ensure the applicability of pQCD calculations. A summary of the model input parameters and the variations considered for each one, to evaluate the model uncertainties, is presented in Table 6.1.

This QCD analysis follows the strategy of the earlier CMS QCD analyses [3, 5], and the parametrization of the PDFs is derived from the most recent one [5], which

Table 6.1 Initial values and variations of the model input parameters used in the QCD analysis. The strong coupling constant and quark masses are treated according to the $\overline{\text{MS}}$-scheme at NLO and the masses listed for m_c and m_b correspond to their pole masses at NLO

Parameter	Initial value	Variation
α_S	0.118	–
m_c	1.5 GeV	$1.37 \leq m_c \leq 1.55$ GeV
m_b	4.5 GeV	$4.3 \leq m_b \leq 5.0$ GeV
Q_0^2	1.9 GeV	$1.6 \leq Q_0^2 \leq 2.2$ GeV
$Q_{\min}^2$	3.5 GeV	$2.5 \leq Q_{\min}^2 \leq 5.0$ GeV

includes the CMS 8 TeV $W^{\pm}$-asymmetry in the fit. The initial scale of the QCD evolution is set to $\mu_0^2 = 1.9\,\text{GeV}^2$. At this scale, the PDFs of the gluon distribution (xg), the valence quark distributions (xu_v, xd_v) and the sea quark distributions, with the u-type $(x\bar{u})$ and d-type $(x\bar{d}, x\bar{s})$ distributions are parametrized as presented in the Eqs. 6.2–6.7, with xs $(x\bar{s})$ denoting the strange quark distribution, and it is assumed that $xs = x\bar{s}$.

$$xu_v(x) = A_{u_v}\, x^{B_{u_v}}\, (1-x)^{C_{u_v}}\, (1 + E_{u_v} x^2), \tag{6.2}$$

$$xd_v(x) = A_{d_v}\, x^{B_{d_v}}\, (1-x)^{C_{d_v}}, \tag{6.3}$$

$$x\bar{u}(x) = A_{\bar{u}}\, x^{B_{\bar{u}}}\, (1-x)^{C_{\bar{u}}}\, (1 + E_{\bar{u}} x^2), \tag{6.4}$$

$$x\bar{d}(x) = A_{\bar{d}}\, x^{B_{\bar{d}}}\, (1-x)^{C_{\bar{d}}}, \tag{6.5}$$

$$x\bar{s}(x) = A_{\bar{s}}\, x^{B_{\bar{s}}}\, (1-x)^{C_{\bar{s}}}, \tag{6.6}$$

$$xg(x) = A_g\, x^{B_g}\, (1-x)^{C_g}. \tag{6.7}$$

The parameters A_{u_v}, A_{d_v} and A_g ensure the normalization of the PDFs according to the QCD sum rules. Therefore constraining $A_{\bar{u}} = A_{\bar{d}}$ regulates the behaviour of the $\bar{u}$ and $\bar{d}$ densities towards $x \to 0$ and ensures that both have the same normalization in this region. In QCD analyses focusing on the extraction of distributions other than strangeness, such as gluons or the valence quarks, the strangeness fraction, defined as $f_s = x\bar{s}/(x\bar{d} + x\bar{s})$ is usually fixed at a constant value to connect the A parameters. However, since this analysis aims to extract the strange quark content of the proton, f_s is a free parameter with no influence on the results. No restrictions are placed on the B parameters of the light sea, as is suggested in Ref. [18].

6.2   Details and Results of the Fit

The global χ^2 of the fitted PDFs is calculated using the *Nuisance Parameter* representation of the correlated and uncorrelated uncertainties associated with the measurements (see Sect. 2.4.2). The initial parameters of the PDFs are determined by

Table 6.2 The partial χ^2 per number of data points for each dataset used in the QCD analysis and the global χ^2 per number of degrees of freedom (N_{dof}) determined from the PDF fit

Data set		χ^2/n_{dp}
HERA I+II charged current	e^+p	43/39
HERA I+II charged current	e^-p	57/42
HERA I+II neutral current	e^-p	218/159
HERA I+II neutral current	e^+p, $E_p = 820\,\text{GeV}$	69/70
HERA I+II neutral current	e^+p, $E_p = 920\,\text{GeV}$	448/377
HERA I+II neutral current	e^+p, $E_p = 460\,\text{GeV}$	216/204
HERA I+II neutral current	e^+p, $E_p = 575\,\text{GeV}$	220/254
CMS $W^\pm$ muon charge asymmetry 7 TeV		13/11
CMS $W^\pm$ muon charge asymmetry 8 TeV		4.2/11
CMS $W + c$ 7 TeV		2.2/5
CMS $W + c$ 13 TeV		2.1/5
Correlated χ^2		87
Total χ^2/dof		1385/1160

minimizing χ^2 and the quality of the fit can be assessed by dividing χ_0^2 by the number of degrees of freedom N_{dof}. Similarly, the partial χ^2 of the individual datasets, divided by the number of data points within the set is determined. The global and partial χ^2 of each dataset is listed in Table 6.2, and good agreement between the CMS measurements is observed. The high χ^2/N_{dof} values determined for the HERA datasets have been investigated in [4]. It was found that the low-Q^2 and medium-Q^2 measurements show some tension, while the high-Q^2 data cannot be fitted well in general. However, it was not possible to pin down the tension to a particular region of Bjorken x or a specific process.

The PDF uncertainties resulting from the experimental uncertainties associated with the input datasets are evaluated using the Hessian method and the MC replica method (see Sect. 2.4.2). When using the Hessian method to estimate the experimental uncertainties, a tolerance criterion of $\Delta\chi^2 = 1$ is applied, corresponding to 68% confidence level. For the error estimation via the MC replica method, a total of 426 replicas of pseudo-data are generated and the uncertainties are determined as the root-mean-square (RMS). Figure 6.3 presents the strange quark content $s(x, \mu^2)$ and the strangeness suppression factor $r_s(x, \mu^2)$ evaluated at the starting scale $\mu_0^2 = 1.9\,\text{GeV}^2$ and the $W^\pm$ boson mass $\mu^2 = m_W^2$. A larger uncertainty over the entire range of Bjorken-x is observed for the use of the MC replica method, indicating non-Gaussian tails in the PDF uncertainties.

The impact of model uncertainties are estimated according to the parameter variations listed in Table 6.1. The PDFs resulting from the fits, using the altered parameters, are compared to the central distributions. No significant differences, with regard to

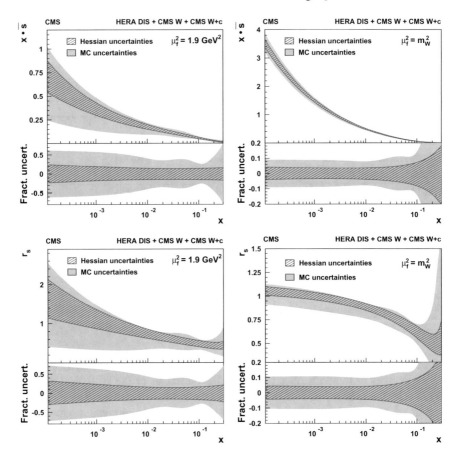

Fig. 6.3 The strange quark distribution (upper) and strangeness suppression factor (lower) as a function of x, obtained at the starting scale $\mu_f^2 = 1.9\,\text{GeV}^2$ (left) and $\mu_f^2 = m_W^2$ (right). The PDF uncertainties, resulting from the experimental uncertainties of the input datasets are evaluated using the Hessian method (hatched band) and the MC replica method (shaded band)

the PDF uncertainties, are observed. Comparisons between the original fit, including the PDF uncertainties, with the results using the parameter variations are presented in Appendix D. Again, no significant deviations from the central result is observed.

6.3 Comparisons with Results of Global PDFs

Here, the results of the QCD analysis, in particular the strangeness suppression r_s, is compared to the PDFs obtained in global fits. In Fig. 6.4, the sets ABMP16nlo, CT14nlo, and MMHT14nlo are considered. The distribution of the strangeness sup-

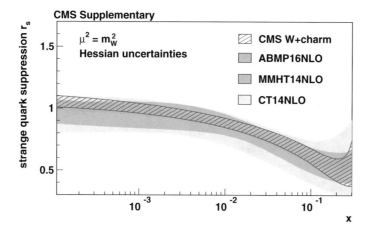

Fig. 6.4 The strangeness suppression r_s as a function of x obtained at the scale $\mu_f^2 = m_W^2$. The results of the analysis presented in this thesis (shaded band) are compared to the distributions of the ABMP16nlo (grey band), CT14nlo (yellow band) and MMHT14nlo (green band) PDF sets

pression, as determined in this thesis is in good agreement, within uncertainties, with the results of the global PDF groups, where the information on the strange quark content in the proton is obtained using the data of neutrino scattering experiments.

In the global PDF fits, additional constraints on the isospin asymmetry (Eq. 6.8) are imposed by using the DY measurement of the NuSea collaboration [19].

$$I(x, \mu^2) = \left[\bar{d}(x, \mu^2) + \bar{u}(x, \mu^2)\right] / \left[\bar{u}(x, \mu^2) + \bar{d}(x, \mu^2)\right] \qquad (6.8)$$

It is worthwhile to note that the results of this thesis are in good agreement with the ABMP16nlo PDF, which uses the most precise NOMAD [20] measurements to probe the strange quark content via neutrino scattering. Therefore it can be concluded that the CMS measurements are in very good agreement with those obtained in neutrino scattering experiments. Figure 6.5 presents a comparison between the strangeness suppression determined in this analysis with the results of ABMP16nlo and ATLASepWZ16nnlo, obtained at the scales $\mu_f^2 = 1.9\,\text{GeV}^2$ and $\mu_f^2 = m_W^2$. For this qualitative comparison only the Hessian uncertainties of ATLASepWZ16nnlo are shown. The unsuppressed strange quark content of the proton, reported by this analysis is not observed. A direct comparison of the results of this thesis to the strange quark distribution and strangeness suppression of ATLASepWZWjet19nnlo [21] was not possible as the necessary files have not been made public as of yet.

However, a comparison of the results obtained in this thesis with the NuSea [19] DY measurements do not show good agreement with each other, either. The reason for this disagreement is that the datasets included in this QCD analysis do not essentially constrain the $\bar{u}$ and $\bar{d}$ distributions. In general, using a limited number of datasets brings the advantage of using a low value for the tolerance criterion, however, this approach suffers from rigidities in the parametrization at the starting scale. The latter

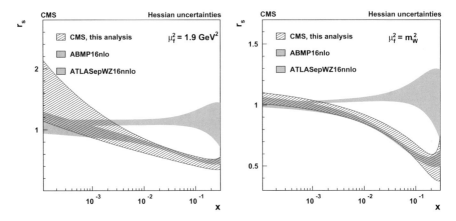

Fig. 6.5 The strangeness suppression factor r_s (Eq. 2.43) as a function of x obtained at the starting scale $\mu_f^2 = 1.9\,\text{GeV}^2$ (left) and $\mu_f^2 = m_W^2$ (right). The results of the analysis presented in this thesis (hatched band) are compared to the distributions of the ABMP16nlo (dark shaded band) and ATLASepWZ16nnlo (light shaded band) PDF sets

can introduce artificial constraints on the PDFs that are not probed directly by the measurements used in the fit. This may lead to effects like an enhanced strange quark distribution i.e. at the cost of an unphysical isospin asymmetry. The QCD analysis presented in this thesis is an illustration of the constraining power of the W + c measurement on the strange quark distribution. The measurement itself [22] is provided in a format suitable to be used by the PDF groups for future global PDF fits.

6.4   Comparison with Previous CMS Results

The parametrization of the 7 TeV CMS QCD analysis [3] differs from the one in this thesis in terms of the used parametrization. For a suitable comparison between the two analyses, the same parametrization is used in a QCD fit with the input datasets listed in Sect. 6.1. The model input parameters, listed in Table 6.1 remain unchanged. The analysis uses a so called *free-s* fit with 15 parameters, where a flexible form of the gluon was adopted, which allows the distribution to assume negative values:

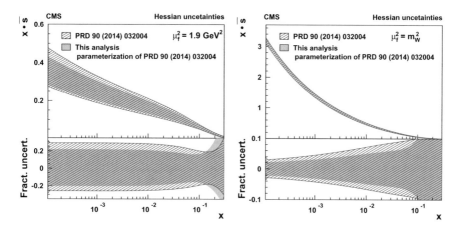

Fig. 6.6 Strange quark distribution in the proton and relative uncertainty of the PDFs (lower panel) as a function of x, evaluated at the starting scale $\mu^2 = 1.9\,\text{GeV}^2$ (left) and $\mu^2 = m_W^2$ (right). The result of the analysis presented in this thesis (filled band) is compared to the results of a previous CMS QCD analysis [3]

$$xu_v(x) = A_{u_v}\, x^{B_{u_v}}\, (1-x)^{C_{u_v}}\, (1 + E_{u_v} x^2), \tag{6.9}$$

$$xd_v(x) = A_{d_v}\, x^{B_{d_v}}\, (1-x)^{C_{d_v}}, \tag{6.10}$$

$$x\bar{u}(x) = A_{\bar{u}}\, x^{B_{\bar{u}}}\, (1-x)^{C_{\bar{u}}}, \tag{6.11}$$

$$x\bar{d}(x) = A_{\bar{d}}\, x^{B_{\bar{d}}}\, (1-x)^{C_{\bar{d}}}, \tag{6.12}$$

$$x\bar{s}(x) = A_{\bar{s}}\, x^{B_{\bar{s}}}\, (1-x)^{C_{\bar{s}}}, \tag{6.13}$$

$$xg(x) = A_g\, x^{B_g}\, (1-x)^{C_g} + A_g'\, x^{B_g'}\, (1-x)^{C_g'}. \tag{6.14}$$

The constraints $A_{\bar{u}} = A_{\bar{d}}$ and $B_{\bar{u}} = B_{\bar{d}}$ as well as $B_{\bar{d}} = B_{\bar{s}}$ are applied, and it is assumed that $xs = x\bar{s}$. These restrictions were necessary at the time due to the limited number of available measurements. The constraint of $B_{\bar{u}} = B_{\bar{d}}$ was released to estimate the parametrization uncertainty, which has a large impact on the total PDF uncertainty. Figure 6.6 presents a comparison of the previous CMS QCD analysis [3] with the results obtained in this thesis. Good agreement between the central values and a reduction of the uncertainties associated with the strange quark distribution is observed when the results of the 13 TeV W + c measurement are included in the fit.

References

1. Alekhin S et al (2015) HERAFitter. Eur Phys J C 75:304. https://doi.org/10.1140/epjc/s10052-015-3480-z. arXiv:1410.4412 [hep-ph]
2. xFitter web site, http://www.xfitter.org/xFitter (2018)

3. Chatrchyan S et al (2014) Measurement of the muon charge asymmetry in inclusive pp → W+X production at $\sqrt{s}$ = 7 TeV and an improved determination of light parton distribution functions. Phys Rev D 90:032004. https://doi.org/10.1103/PhysRevD.90.032004. arXiv:1312.6283 [hep-ex]
4. H1 and ZEUS Collaborations (2015) Combination of measurements of inclusive deep inelastic $e^{\pm}$p scattering cross sections and QCD analysis of HERA data. Eur Phys J C 75:580. https://doi.org/10.1140/epjc/s10052-015-3710-4. arXiv:1506.06042 [hep-ex]
5. Khachatryan V et al (2016) Measurement of the differential cross section and charge asymmetry for inclusive pp → $W^{\pm}$ + X production at $\sqrt{s}$ = 8 TeV. Eur Phys J C 76:469. https://doi.org/10.1140/epjc/s10052-016-4293-4. arXiv:1603.01803 [hep-ex]
6. Li Y, Petriello F (2012) Combining QCD and electroweak corrections to dilepton production in FEWZ. Phys Rev D 86:094034. https://doi.org/10.1103/PhysRevD.86.094034. arXiv:1208.5967 [hep-ph]
7. Chatrchyan S et al (2014) Measurement of associated W+charm production in pp collisions at $\sqrt{s}$ = 7 TeV. JHEP 02:013. https://doi.org/10.1007/JHEP02(2014)013. arXiv:1310.1138 [hep-ex]
8. Campbell JM, Ellis RK (2010) MCFM for the Tevatron and the LHC. Nucl Phys Proc Suppl 205–206:10. https://doi.org/10.1016/j.nuclphysbps.2010.08.011. arXiv: 1007.3492 [hep-ph]
9. Campbell JM, Keith Ellis R (1999) An update on vector boson pair production at hadron colliders. Phys Rev D 60:113006. https://doi.org/10.1103/PhysRevD.60.113006. arXiv: hep--ph/9905386 [hep-ph]
10. Campbell JM, Keith Ellis R (2015) Top quark processes at NLO in production and decay. J Phys G 42:015005. https://doi.org/10.1088/0954-3899/42/1/015005. arXiv:1204.1513 [hep-ph]
11. Carli T et al (2010) A posteriori inclusion of parton density functions in NLO QCD final-state calculations at hadron colliders: the APPLGRID project. Eur Phys J C 66:503. https://doi.org/10.1140/epjc/s10052-010-1255-0. arXiv:0911.2985 [hep-ph]
12. Publicly available experimental data sets in the xFitter package (2019) https://xfitter.hepforge.org/data.html
13. Botje M (2011) QCDNUM: fast QCD evolution and convolution. Comput Phys Commun 182:490. https://doi.org/10.1016/j.cpc.2010.10.020. arXiv:1005.1481 [hep-ph]
14. Gerard 't Hooft (1973) Dimensional regularization and the renormalization group. Nucl Phys B61:455-468. https://doi.org/10.1016/0550-3213(73)90376-3
15. Bardeen WA et al (1978) Deep inelastic scattering beyond the leading order in asymptotically free gauge theories. Phys Rev D 18:3998. https://doi.org/10.1103/PhysRevD.18.3998
16. Thorne RS (2006) Variable-flavor number scheme for NNLO. Phys Rev D 73:054019. https://doi.org/10.1103/PhysRevD.73.054019. arXiv:hepph/0601245 [hep-ph]
17. Martin AD et al (2009) Parton distributions for the LHC. Eur Phys J C 63:189. https://doi.org/10.1140/epjc/s10052-009-1072-5. arXiv:0901.0002 [hep-ph]
18. Alekhin S, Blümlein J, Moch S (2018) Strange sea determination from collider data. Phys Lett B 777:134–140. https://doi.org/10.1016/j.physletb.2017.12.024. arXiv:1708.01067 [hep-ph]
19. Towell RS et al (2001) Improved measurement of the anti-d / anti-u asymmetry in the nucleon sea. Phys Rev D 64:052002. https://doi.org/10.1103/PhysRevD.64.052002. arXiv:hep-ex/0103030 [hep-ex]
20. Samoylov O et al (2013) A precision measurement of charm dimuon production in neutrino interactions from the NOMAD experiment. Nucl Phys B 876:339. https://doi.org/10.1016/j.nuclphysb.2013.08.021. arXiv:1308.4750 [hep-ex]
21. QCD analysis of ATLAS $W^{\pm}$ boson production data in association with jets. Technical report ATL-PHYS-PUB-2019-016. Geneva: CERN (2019). https://cds.cern.ch/record/2670662
22. Sirunyan AM et al (2019) Measurement of associated production of a W boson and a charm quark in proton-proton collisions at $\sqrt{s}$ = 13 TeV. Eur Phys J C 79(3):269. https://doi.org/10.1140/epjc/s10052-019-6752-1. arXiv:1811.10021 [hep-ex]

Chapter 7
Summary and Conclusions

In this thesis, the measurement of the associated production of a $W^{\pm}$ boson and a charm quark in 13 TeV pp-collisions at the LHC is described. The impact of these results in improving constraints on the strange quark content of the proton is investigated in a QCD analysis at next-to-leading order, using the XFITTER framework.

The data used in the W + c measurement was recorded by the CMS detector in 2016 and corresponds to an integrated luminosity of 35.7 fb^{-1}. The $W^{\pm}$ bosons are reconstructed via their leptonic decay $W^{\pm} \rightarrow \mu\nu$, which has a clear event signature with a single, isolated, high-p_T muon and missing transverse momentum $\vec{p}_T^{\text{miss}}$, indicating the presence of a neutrino. The charm quarks are tagged by the full reconstruction of charmed meson decays $D^*(2010)^{\pm} \rightarrow D^0 + \pi^{\pm}_{\text{slow}} \rightarrow K^{\mp} + \pi^{\pm} + \pi^{\pm}_{\text{slow}}$, resulting in a clear peak around the expected mass difference of $\Delta m(D^0, D^*) = 0.1454$ GeV. Contributions from gluon splitting background processes (W + c$\bar{\text{c}}$, W + b$\bar{\text{b}}$), are removed using a data driven method that is considering the fact that the $W^{\pm}$ boson and the charm quark produced in W + c always have opposite charge signs, whereas the contributions from gluon splitting have the same probabilities for W + c combinations of the same and opposite charge. The acceptance and efficiency of event selection is determined via a Monte Carlo simulation of W + $D^*(2010)^{\pm}$, generated with MADGRAPH5_aMC@NLO [8] and interfaced with PYTHIA (V8.2.12) [12] for parton shower and hadronization.

The fiducial W + c cross section is determined in a kinematic range defined by the transverse momentum of the muon $p_T^{\mu} > 26$ GeV, the pseudorapidity $|\eta^{\mu}| < 2.4$ and the charm transverse momentum $p_T^c > 5$ GeV. The measurements are performed inclusively and in five bins of $|\eta^{\mu}|$, with the available Monte Carlo statistics and the fragmentation of $c \rightarrow D^*(2010)^{\pm}$ as the dominant systematic uncertainties. The obtained values for the inclusive fiducial W + c cross section and for the cross section ratio are:

S. K. Pflitsch, *Associated Production of W + Charm in 13 TeV*
Proton-Proton Collisions Measured with CMS and Determination of the Strange Quark
Content of the Proton, Springer Theses,
https://doi.org/10.1007/978-3-030-52762-4_7

$$\sigma(W + c) = 1026 \pm 31 \text{ stat.} {}^{+76}_{-72} \text{syst. pb} \tag{7.1}$$

$$\frac{\sigma(W^+ + \bar{c})}{\sigma(W^- + c)} = 0.968 \pm 0.055 \text{ stat.} {}^{+0.015}_{-0.028} \text{syst.} \tag{7.2}$$

The inclusive and differential measurements are compared to theoretical predictions at next-to-leading order of QCD, obtained using the MCFM [9–11] program in combination with different sets of parton distribution functions. Good agreement between the measurements and the theoretical predictions is observed for all calculations using global the PDF sets ABMP16nlo [3], CT14nlo [4], MMHT14nlo [5] and NNPDF3.1nlo [6].

The differential results of the $W + c$ measurement are sensitive to the strange quark content of the proton in the range of $10^{-3} \le x \le 10^{-1}$ and are therefore used in a subsequent QCD analysis to determine the strange quark distribution and the strangeness suppression $r_s = (s + \bar{s})/(\bar{u} + \bar{d})$. The fit is performed using the combination of the HERA I+II [1] datasets, the CMS measurements of the $W^{\pm}$ asymmetry at 7 and 8 TeV and the CMS $W + c$ measurements at 7 and 13 TeV. The experimental uncertainties associated with the input datasets are propagated to the PDF uncertainties using the Hessian method with a tolerance criterion of $\Delta\chi^2 = 1$, and the MC replica method for comparison. Both results are in agreement with each other, though the uncertainties obtained via the MC replica method is larger than what is observed when the Hessian method is applied. Uncertainties arising from assumptions on the model input parameters, such as the heavy quark masses or the value of $Q^2_{\min}$ imposed on the HERA data were investigated but found to be negligible in this analysis. The resulting distribution of r_s is compatible with those of the global PDF sets and disagrees with ATLASepWZ16nnlo [7], for which an unsuppressed strangeness is reported.

Further, the measurements are used in an alternative QCD analysis, utilizing the parametrization of the previous CMS result [2] and are found to be in good agreement. Therefore, the measurement of $W + c$ production in proton-proton collisions at the LHC demonstrates that it provides important information about the flavour composition of the proton quark sea and is in agreement with the results of neutrino scattering experiments.

The analysis of this thesis has recently been published in *The European Physical Journal C* [13] and the measurements are provided in a form suitable to be included in the global PDF fits along with the corresponding theory predictions.

References

1. H1 and ZEUS Collaborations (2015) Combination of measurements of inclusive deep inelastic e^{pm}p scattering cross sections and QCD analysis of HERA data. Eur Phys J C 75:580. https://doi.org/10.1140/epjc/s10052-015-3710-4. arXiv:1506.06042 [hep-ex]
2. Chatrchyan S et al (2014) Measurement of the muon charge asymmetry in inclusive pp → W+X production at $\sqrt{s} = 7$ TeV and an improved determination of light parton distribution functions.

Phys Rev D 90:032004. https://doi.org/10.1103/PhysRevD.90.032004. arXiv:1312.6283 [hep-ex]

3. Alekhin S, Blümlein J, Moch S (2018) NLO PDFs from the ABMP16 fit. Eur Phys J C 78:477. https://doi.org/10.1140/epjc/s10052-018-5947-1. arXiv:1803.07537 [hep-ph]

4. Dulat S et al (2016) New parton distribution functions from a global analysis of quantum chromodynamics. Phys Rev D 93:033006. https://doi.org/10.1103/PhysRevD.93.033006. arXiv:1506.07443 [hep-ph]

5. Harland-Lang LA et al (2015) Parton distributions in the LHC era: MMHT 2014 PDFs. Eur Phys J C 75:204. https://doi.org/10.1140/epjc/s10052-015-3397-6. arXiv:1412.3989 [hep-ph]

6. Ball RD et al (2017) Parton distributions from high-precision collider data. Eur Phys J C 77:663. https://doi.org/10.1140/epjc/s10052-017-5199-5. arXiv:1706.00428 [hep-ph]

7. Aaboud M et al (2017) Precision measurement and interpretation of inclusive W^+, W^- and Z/γ^* production cross sections with the ATLAS detector. Eur Phys J C 77:367. https://doi.org/10.1140/epjc/s10052-017-4911-9. arXiv:1612.03016 [hep-ex]

8. Alwall J et al (2014) The automated computation of tree-level and next-to-leading order differential cross sections, and their matching to parton shower simulations. JHEP 07:079. https://doi.org/10.1007/JHEP07(2014)079. arXiv:1405.0301 [hep-ph]

9. Campbell JM, Ellis RK (2010) MCFM for the Tevatron and the LHC. Nucl Phys Proc Suppl 205–206:10. https://doi.org/10.1016/j.nuclphysbps.2010.08.011. arXiv:1007.3492 [hep-ph]

10. Campbell JM, Keith Ellis R (1999) An update on vector boson pair production at hadron colliders. Phys Rev D 60:113006. https://doi.org/10.1103/PhysRevD.60.113006. arXiv:hep-ph/9905386 [hep-ph]

11. Campbell JM, Keith Ellis R (2015) Top quark processes at NLO in production and decay. J Phys G 42:015005. https://doi.org/10.1088/0954-3899/42/1/015005. arXiv:1204.1513 [hep-ph]

12. Sjöstrand T, Mrenna S, Skands PZ (2008) A brief introduction to PYTHIA 8.1. Comput Phys Commun 178:852. https://doi.org/10.1016/j.cpc.2008.01.036. arXiv: 0710.3820 [hep-ph]

13. Sirunyan AM et al (2019) Measurement of associated production of a W boson and a charm quark in proton-proton collisions at $\sqrt{s}$ = 13 TeV. Eur Phys J C 79(3):269. https://doi.org/10.1140/epjc/s10052-019-6752-1. arXiv:1811.10021 [hep-ex]

Appendix A
W + Charm: Input Datasets

In this analysis, the data collected by the CMS experiment in 2016 is used. Run A is excluded as the magnet had been switched off in this period. To ensure good functionality of all the detector components only the Lumi-Sections listed in the certification-file, presented in Table A.1, are used. The recorded and certified data of the periods B-H correspond to an integrated luminosity of $35.7 \, \text{fb}^{-1}$ [1]. The samples used in the W + c analysis have passed the single muon trigger selections (see Sect. 4.3.1.2).

The Monte Carlo (MC) samples used in this analysis are listed in Table A.2 along with their corresponding cross sections, and the numbers of generated events.

© The Editor(s) (if applicable) and The Author(s), under exclusive license to Springer Nature Switzerland AG 2020
S. K. Pflitsch, *Associated Production of W + Charm in 13 TeV Proton-Proton Collisions Measured with CMS and Determination of the Strange Quark Content of the Proton*, Springer Theses, https://doi.org/10.1007/978-3-030-52762-4

Table A.1 Datasets and certification-file used in the analysis

Dataset name	#Events	Lum.(fb^{-1})
/SingleMuon/Run2016B-07Aug17_ver2-v1	158145722	5.74
/SingleMuon/Run2016C-07Aug17-v1	67441308	2.57
/SingleMuon/Run2016D-07Aug17-v1	98017996	4.25
/SingleMuon/Run2016E-07Aug17-v1	90984718	3.95
/SingleMuon/Run2016F-07Aug17-v2	65489554	3.10
/SingleMuon/Run2016G-07Aug17-v1	149916849	7.50
/SingleMuon/Run2016H-07Aug17-v1	174035164	8.58
JSON		
Cert_271036-284044_13TeV_PromptReco_Collisions16_JSON.txt		

Table A.2 Simulated samples used in the analysis with their respective cross-sections and numbers of generated events

MC-Sample name	#Events	$\sigma(pb)$
Signal		
/WJetsToLNu_DStarFilter_TuneCUETP8M1_13TeV-amcatnloFXFX-pythia8	45 M	2012
W+jets		
/WJetsToLN_TuneCUETP8M1_13TeV-amcatnloFXFX-pythia8	200 M	60320
Other backgrounds		
/DYJetsToLL_M-50_TuneCUETP8M1_13TeV- amcatnloFXFX-pythia8	50 M	6025
/QCD_Pt-20to30_MuEnrichedPt5_TuneCUETP8M1_13TeV _pythia8	32 M	$3.16 \cdot 10^6$
/QCD_Pt-30to50_MuEnrichedPt5_TuneCUETP8M1_13TeV _pythia8	30 M	$1.66 \cdot 10^6$
/QCD_Pt-50to80_MuEnrichedPt5_TuneCUETP8M1_13TeV _pythia8	20 M	$4.50 \cdot 10^5$
/QCD_Pt-80to120_MuEnrichedPt5_TuneCUETP8M1_13TeV _pythia8	14 M	$1.07 \cdot 10^5$
/TT_TuneCUETP8M2T4_13TeV-powheg-pythia8	25 M	832
/ST_t-channel_top_4f_inclusiveDecays_13TeV-powhegV2- madspin-pythia8_TuneCUETP8M1/	10 M	136
/ST_t-channel_antitop_4f_inclusiveDecays_13TeV-powhegV2- madspin-pythia8_TuneCUETP8M1	10 M	81.0
/ST_tW_top_5f_inclusiveDecays_13TeV-powheg-pythia8 _TuneCUETP8M1	7 M	35.6
/ST_tW_antitop_5f_inclusiveDecays_13TeV-powheg-pythia8 _TuneCUETP8M1	7 M	35.6
/ST_s-channel_4f_leptonDecays_13TeV-amcatnlo-pythia8 _TuneCUETP8M1	1 M	10.3
/WW_TuneCUETP8M1_13TeV-pythia8	1 M	64.3
/WZ_TuneCUETP8M1_13TeV-pythia8	1 M	23.5
/ZZ_TuneCUETP8M1_13TeV-pythia8	1 M	15.4

Appendix B
Tracking Studies

This section presents the distributions of tracking parameters that were investigated over the course of the W + c analysis, but are not considered as selection criteria for the $D^*(2010)^\pm$ candidates. All tracks presented in the following pass the event and object selections listed in Sects. 5.2.2 and 5.2.3. The tracks associated with a $K^\mp$ or $\pi^\pm$ candidate are presented separately from the tracks associated with a $\pi^\pm_{\text{slow}}$ candidate, due to the different p_T^{track} thresholds.

Figure B.1 presents the track transverse momentum uncertainty $\sigma p_T^{\text{track}}$ and the relative transverse momentum resolution $\sigma p_T^{\text{track}} / p_T^{\text{track}}$, while Fig. B.2 shows the track pseudorapidity and azimuthal angle uncertainties. Figure B.3 presents the χ^2 distribution of the tracks, their number of degrees of freedom (N_{dof}), and the reduced χ^2. The hit multiplicities of the tracks are presented in Fig. B.4, for the pixel layers, the strip layers, and all tracker layers. Furthermore, the average hit multiplicity in the pixel, strip and all tracker layers is shown as a function of η in Fig. B.5.

© The Editor(s) (if applicable) and The Author(s), under exclusive license
to Springer Nature Switzerland AG 2020
S. K. Pflitsch, *Associated Production of W + Charm in 13 TeV
Proton-Proton Collisions Measured with CMS and Determination of the Strange Quark
Content of the Proton*, Springer Theses,
https://doi.org/10.1007/978-3-030-52762-4

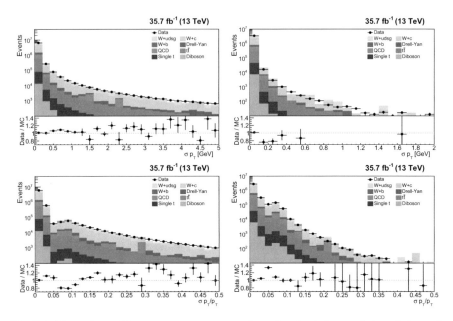

Fig. B.1 Uncertainty of the transverse momentum (upper) and relative resolution of the track transverse momentum for $K^{\mp}$ and $\pi^{\pm}$ candidates (left) and $\pi^{\pm}_{slow}$ candidates (right) in the events fulfilling all selection criteria. The data (filled circles) are compared to the MC simulation of contributions from different processes (filled bands of different color)

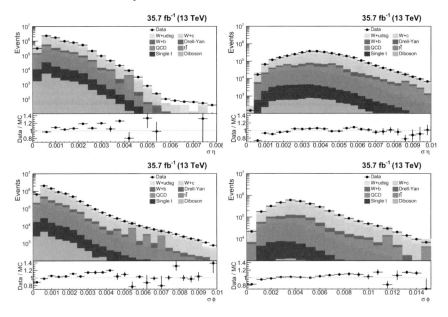

Fig. B.2 Track pseudorapidity (upper) and azimuthal angle (lower) resolution of $K^{\mp}$ and $\pi^{\pm}$ candidates (left) and $\pi^{\pm}_{slow}$ candidates (right) in the events fulfilling all selection criteria. The data (filled circles) are compared to the MC simulation of contributions from different processes (filled bands of different color)

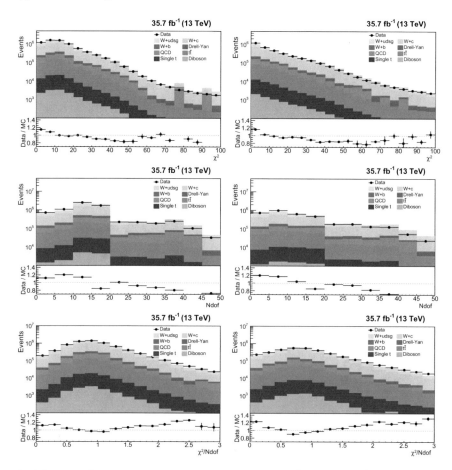

Fig. B.3 Track χ^2, number of degrees of freedom and reduced χ^2 for $K^{\mp}$ and $\pi^{\pm}$ candidates (left) and $\pi^{\pm}_{slow}$ candidates (right) in the events fulfilling all selection criteria. The data (filled circles) are compared to the MC simulation of contributions from different processes (filled bands of different color)

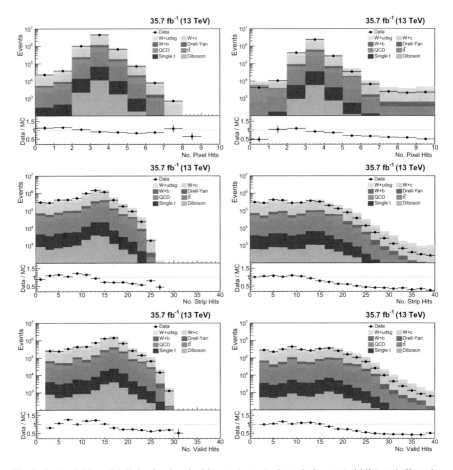

Fig. B.4 Track hit multiplicity for the pixel layers (upper), the strip layers (middle) and all tracker layers (lower) for $K^{\mp}$ and $\pi^{\pm}$ candidates (left) and $\pi^{\pm}_{slow}$ candidates (right) in the events fulfilling all selection criteria. The data (filled circles) are compared to the MC simulation of contributions from different processes (filled bands of different color)

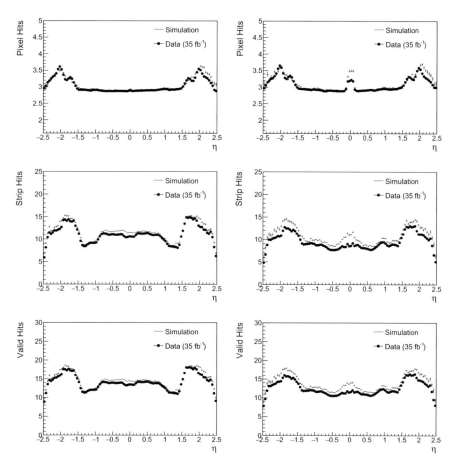

Fig. B.5 Average hit multiplicity in the pixel layers (upper), strip layers (middle) and all tracker layers (lower) as a function of η for $K^{\mp}$ and $\pi^{\pm}$ candidates (left) and $\pi^{\pm}_{slow}$ candidates (right) in the events fulfilling all selection criteria. The data (filled circles) are compared to the MC simulation (blue lines) (Color online)

Appendix C
Binwise D^* Reconstruction

This section presents the distribution of the reconstructed mass difference $\Delta m(\mathrm{D}^0, \mathrm{D}^*)$ at different stages of the W + c analysis for the investigated ranges of $|\eta^\mu|$. All D^0 and $\mathrm{D}^*(2010)^\pm$ candidates fulfill the event and object selections listed in Sects. 5.2.2 and 5.2.3. Figures C.7, C.8, C.9, C.10, C.11, C.12, C.13, C.14, C.15, C.16, C.17, C.18 and C.19 show the distributions for events with $\mathrm{W}^\pm$ candidates of any charge, while Figs. C.14, C.15, C.16, C.17, C.18 and C.19 show the distributions for events with W^+ candidates and Figs. C.19, C.20 and C.21 those for W^- candidates. For each of the $\mathrm{W}^\pm$ charges the plots are presented in the following order (Figs. C.1, C.2, C.3, C.4, C.5 and C.6):

Two figures show the $\Delta m(\mathrm{D}^0, \mathrm{D}^*)$ distribution of the right-charge and wrong-charge candidates, falling into the OS category, in each of the investigated ranges of $|\eta^\mu|$, separately for data and simulation. Additionally two figures presenting the corresponding $\Delta m(\mathrm{D}^0, \mathrm{D}^*)$ distributions for events falling into the SS category are shown. The next two figures present the $\Delta m(\mathrm{D}^0, \mathrm{D}^*)$ distribution after the wrong charge candidates have been subtracted in each of the investigated ranges of $|\eta^\mu|$. Events falling into the OS and SS categories are presented separately. The final figure presents the $\Delta m(\mathrm{D}^0, \mathrm{D}^*)$ distribution after OS $-$ SS.

© The Editor(s) (if applicable) and The Author(s), under exclusive license
to Springer Nature Switzerland AG 2020
S. K. Pflitsch, *Associated Production of W + Charm in 13 TeV
Proton-Proton Collisions Measured with CMS and Determination of the Strange Quark
Content of the Proton*, Springer Theses,
https://doi.org/10.1007/978-3-030-52762-4

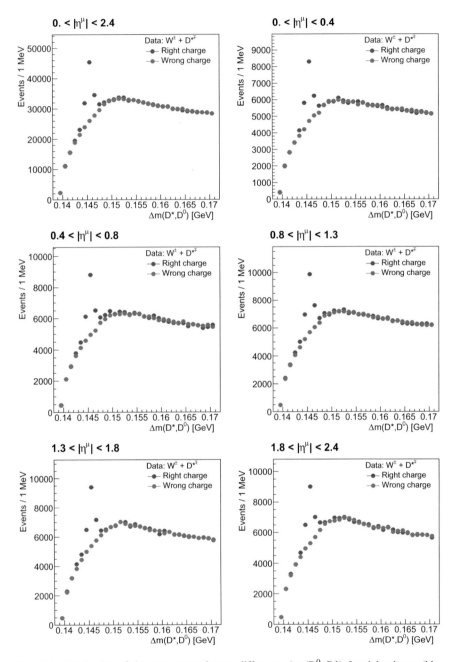

Fig. C.1 Distribution of the reconstructed mass difference $\Delta m(D^0, D^*)$ for right-charge (blue points) and wrong-charge (red points) candidates in $W^\pm + D^*(2010)^\mp$ events after normalization of the wrong-charge distribution for different ranges of $|\eta^\mu|$ in data (Color online)

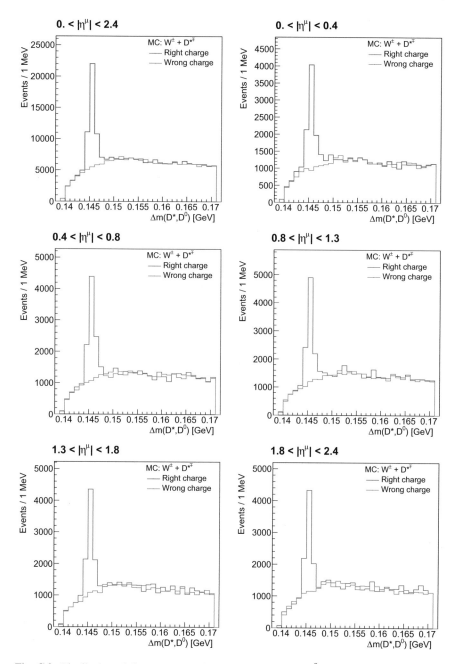

Fig. C.2 Distribution of the reconstructed mass difference $\Delta m(D^0, D^*)$ for right-charge (blue points) and wrong-charge (red points) candidates in $W^\pm + D^*(2010)^\mp$ events after normalization of the wrong-charge distribution for different ranges of $|\eta^\mu|$ in simulation (Color online)

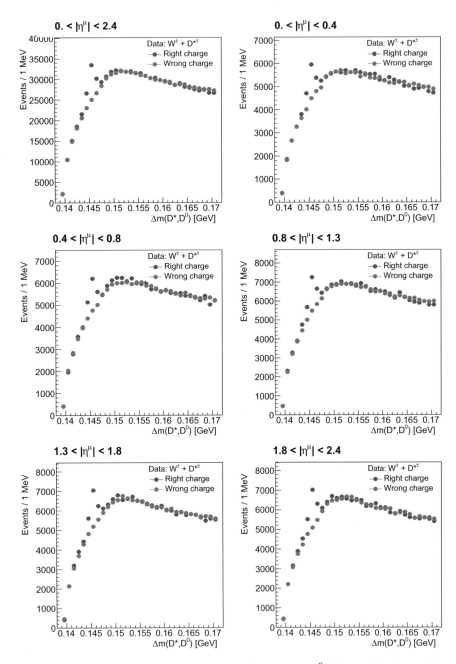

Fig. C.3 Distribution of the reconstructed mass difference $\Delta m(D^0, D^*)$ for right-charge (blue points) and wrong-charge (red points) candidates in $W^\pm + D^*(2010)^\pm$ events after normalization of the wrong-charge distribution for different ranges of $|\eta^\mu|$ in data (Color online)

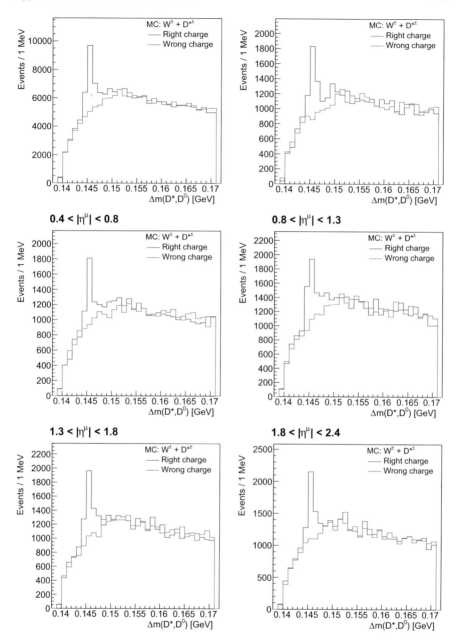

Fig. C.4 Distribution of the reconstructed mass difference $\Delta m(D^0, D^*)$ for right-charge (blue points) and wrong-charge (red points) candidates in $W^{\pm} + D^*(2010)^{\pm}$ events after normalization of the wrong-charge distribution for different ranges of $|\eta^{\mu}|$ in simulation (Color online)

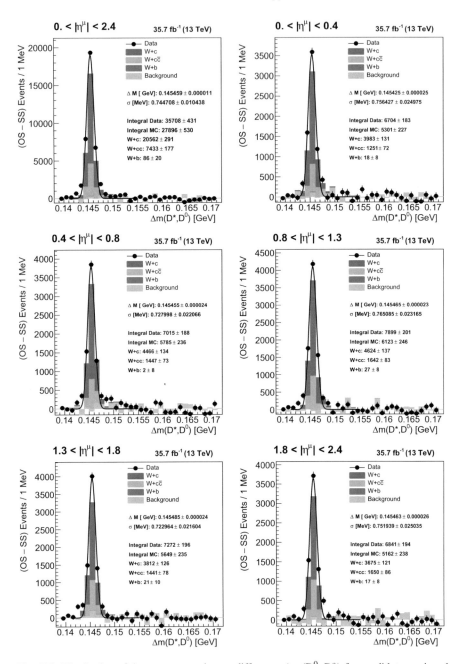

Fig. C.5 Distribution of the reconstructed mass difference $\Delta m(D^0, D^*)$ for candidates assigned to the OS category for different ranges of $|\eta^\mu|$. The combinatorial background has been subtracted and data are compared to MC simulation with contributions from different processes shown as histograms of different colours

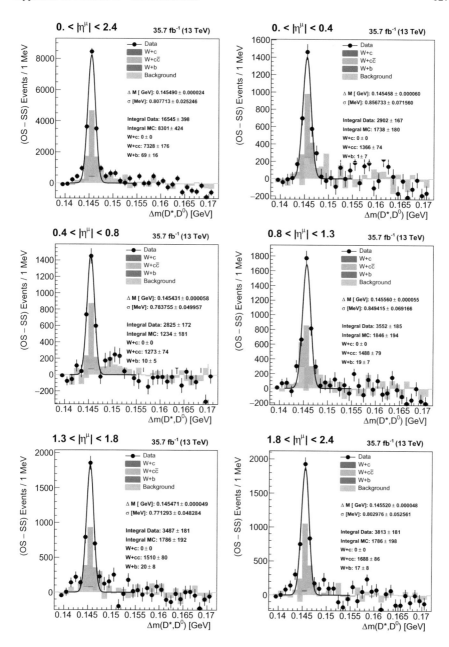

Fig. C.6 Distribution of the reconstructed mass difference $\Delta m(D^0, D^*)$ for candidates assigned to the SS category for different ranges of $|\eta^\mu|$. The combinatorial background has been subtracted and data are compared to MC simulation with contributions from different processes shown as histograms of different colours

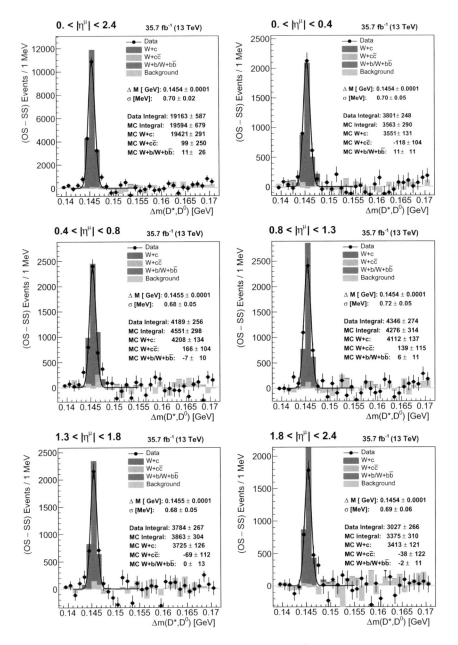

Fig. C.7 Distribution of the reconstructed mass difference $\Delta m(D^0, D^*)$ after OS − SS. The combinatorial background has been subtracted and data are compared to MC simulation with contributions from different processes shown as histograms of different colours

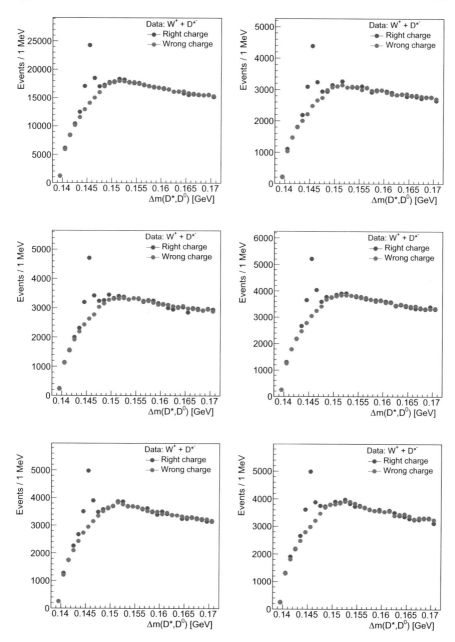

Fig. C.8 Distribution of the reconstructed mass difference $\Delta m(D^0, D^*)$ for right-charge (blue points) and wrong-charge (red points) candidates in $W^+ + D^*(2010)^-$ events after normalization of the wrong-charge distribution for different ranges of $|\eta^\mu|$ in data (Color online)

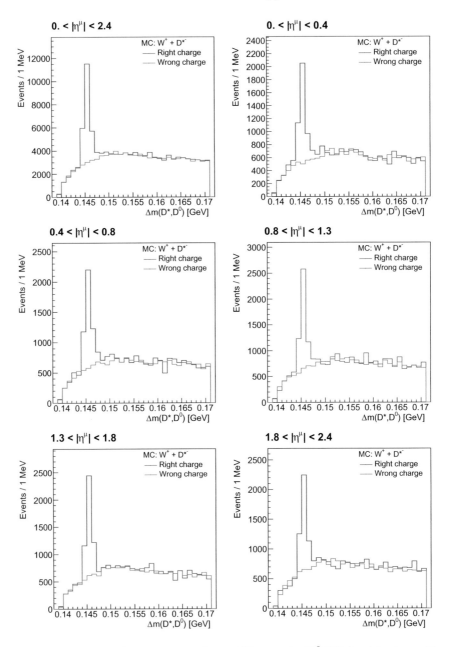

Fig. C.9 Distribution of the reconstructed mass difference $\Delta m(\mathrm{D}^0, \mathrm{D}^*)$ for right-charge (blue line) and wrong-charge (red line) candidates in $\mathrm{W}^+ + \mathrm{D}^*(2010)^-$ events after normalization of the wrong-charge distribution for different ranges of $|\eta^\mu|$ in simulation (Color online)

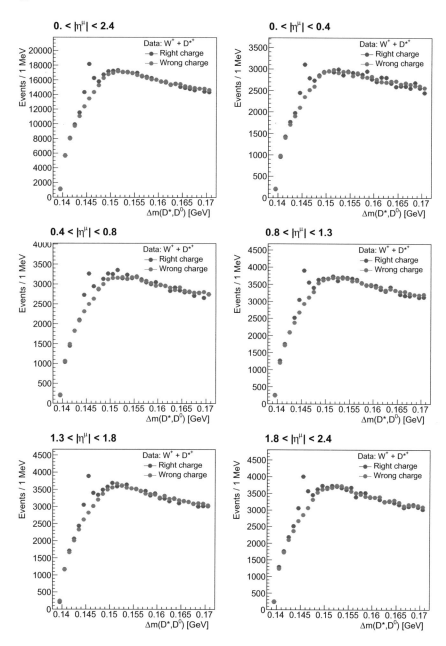

Fig. C.10 Distribution of the reconstructed mass difference $\Delta m(D^0, D^*)$ for right-charge (blue points) and wrong-charge (red points) candidates in $W^+ + D^*(2010)^+$ events after normalization of the wrong-charge distribution for different ranges of $|\eta^\mu|$ in data (Color online)

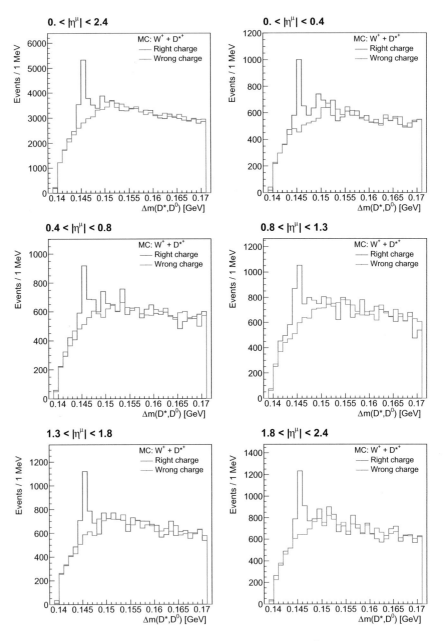

Fig. C.11 Distribution of the reconstructed mass difference $\Delta m(D^0, D^*)$ for right-charge (blue line) and wrong-charge (red line) candidates in $W^+ + D^*(2010)^+$ events after normalization of the wrong-charge distribution for different ranges of $|\eta^\mu|$ in simulation (Color online)

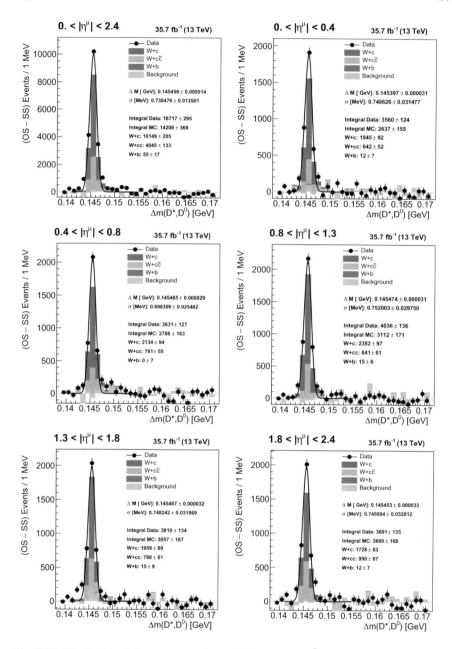

Fig. C.12 Distribution of the reconstructed mass difference $\Delta m(D^0, D^*)$ for candidates assigned to the OS category in events with a W^+ candidate for different ranges of $|\eta^\mu|$. The combinatorial background has been subtracted and data are compared to MC simulation with contributions from different processes shown as histograms of different colours

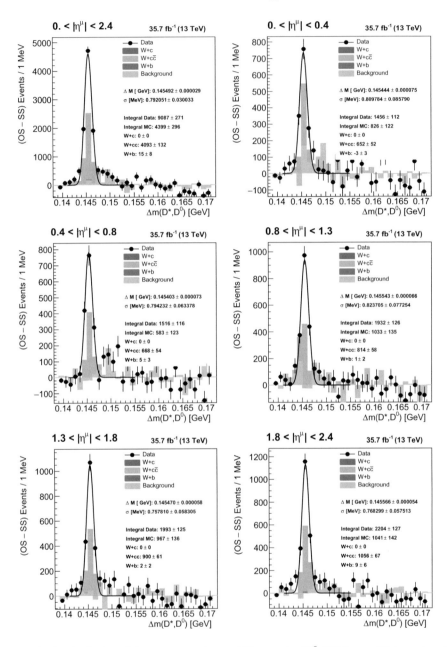

Fig. C.13 Distribution of the reconstructed mass difference $\Delta m(\mathrm{D}^0, \mathrm{D}^*)$ for candidates assigned to the SS category in events with a W^+ candidate for different ranges of $|\eta^\mu|$. The combinatorial background has been subtracted and data are compared to MC simulation with contributions from different processes shown as histograms of different colours

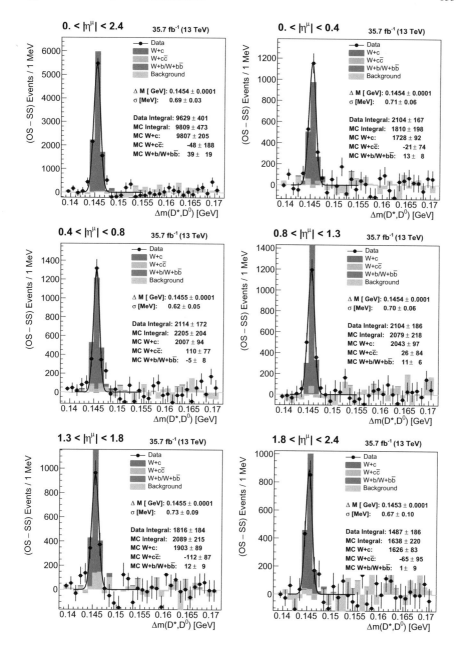

Fig. C.14 Distribution of the reconstructed mass difference $\Delta m(D^0, D^*)$ after OS − SS for events with a W^+ candidate. The combinatorial background has been subtracted and data are compared to MC simulation with contributions from different processes shown as histograms of different colours

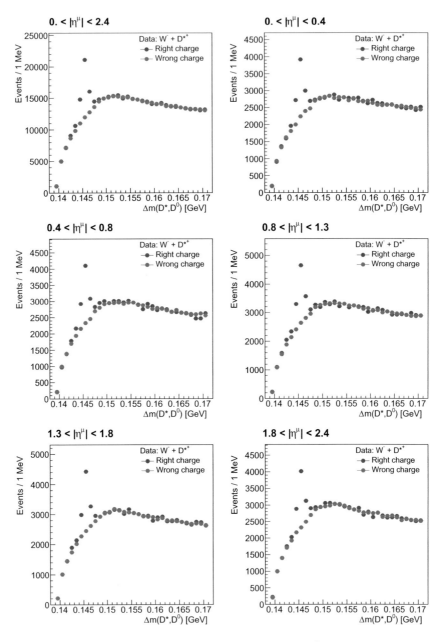

Fig. C.15 Distribution of the reconstructed mass difference $\Delta m(D^0, D^*)$ for right-charge (blue points) and wrong-charge (red points) candidates in $W^- + D^*(2010)^+$ events after normalization of the wrong-charge distribution for different ranges of $|\eta^\mu|$ in data (Color online)

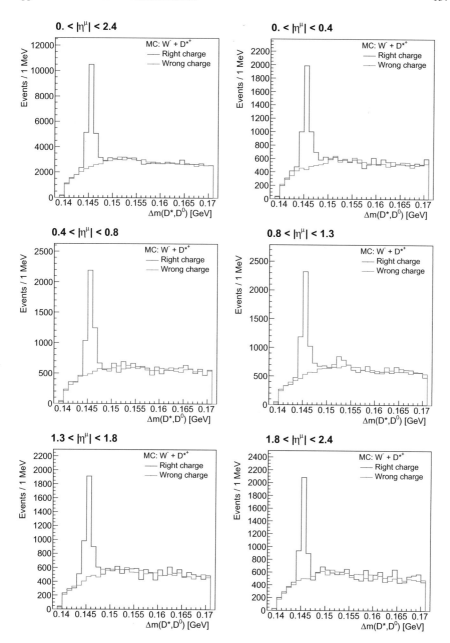

Fig. C.16 Distribution of the reconstructed mass difference $\Delta m(D^0, D^*)$ for right-charge (blue line) and wrong-charge (red line) candidates in $W^- + D^*(2010)^+$ events after normalization of the wrong-charge distribution for different ranges of $|\eta^\mu|$ in simulation (Color online)

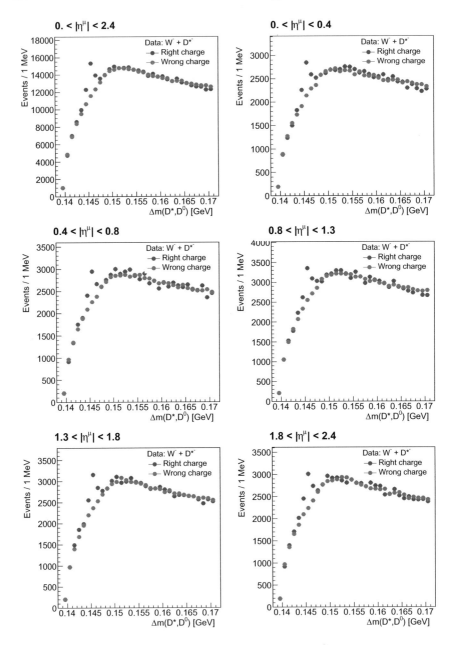

Fig. C.17 Distribution of the reconstructed mass difference $\Delta m(\mathrm{D}^0, \mathrm{D}^*)$ for right-charge (blue points) and wrong-charge (red points) candidates in $\mathrm{W}^- + \mathrm{D}^*(2010)^-$ events after normalization of the wrong-charge distribution for different ranges of $|\eta^\mu|$ in data (Color online)

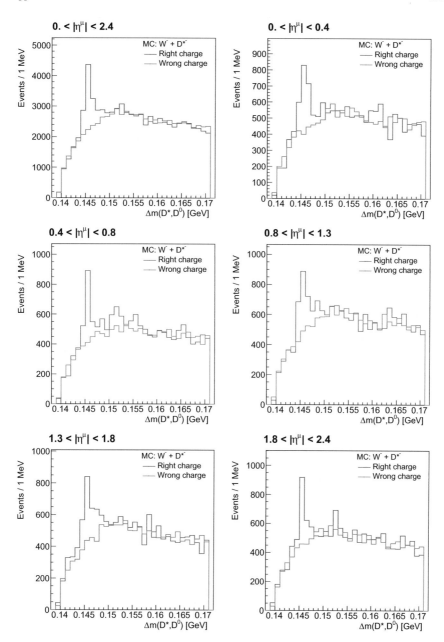

Fig. C.18 Distribution of the reconstructed mass difference $\Delta m(D^0, D^*)$ for right-charge (blue line) and wrong-charge (red line) candidates in $W^- + D^*(2010)^-$ events after normalization of the wrong-charge distribution for different ranges of $|\eta^\mu|$ in simulation (Color online)

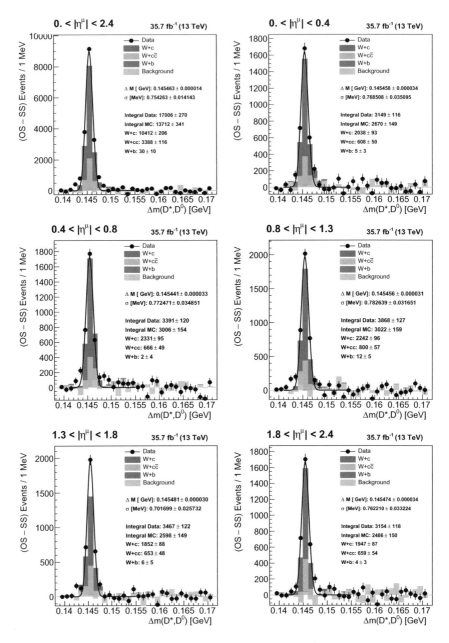

Fig. C.19 Distribution of the reconstructed mass difference $\Delta m(D^0, D^*)$ for candidates assigned to the OS category in events with a W^- candidate for different ranges of $|\eta^\mu|$. The combinatorial background has been subtracted and data are compared to MC simulation with contributions from different processes shown as histograms of different colours

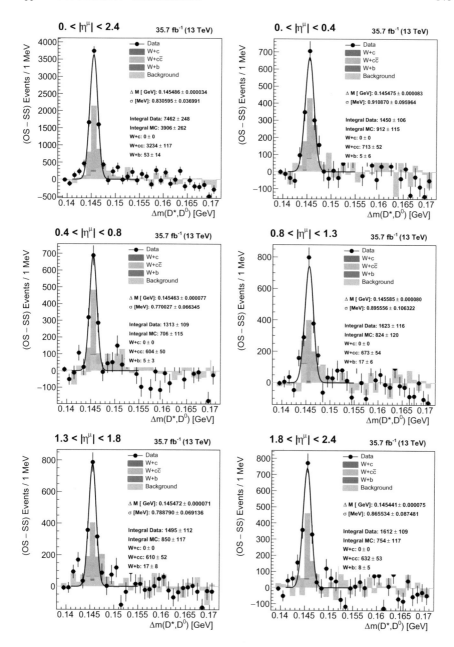

Fig. C.20 Distribution of the reconstructed mass difference $\Delta m(D^0, D^*)$ for candidates assigned to the SS category in events with a W^- candidate for different ranges of $|\eta^\mu|$. The combinatorial background has been subtracted and data are compared to MC simulation with contributions from different processes shown as histograms of different colours

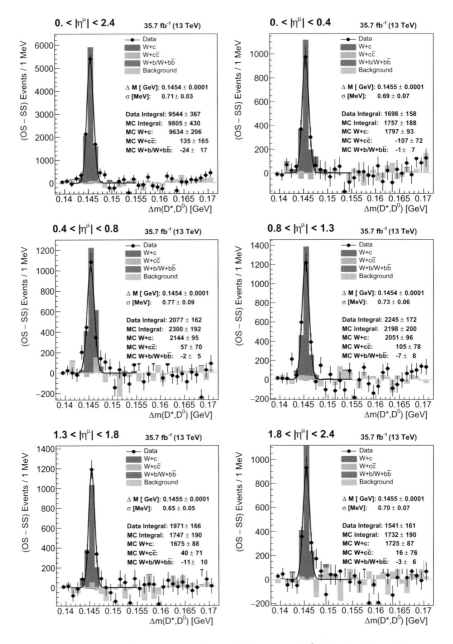

Fig. C.21 Distribution of the reconstructed mass difference $\Delta m(D^0, D^*)$ after OS − SS for events with a W^+ candidate. The combinatorial background has been subtracted and data are compared to MC simulation with contributions from different processes shown as histograms of different colours

Appendix D
QCD Analysis: Model Uncertainties

This section presents the model and parametrization uncertainties, investigated over the course of the QCD analysis. The strangeness distribution and strangeness suppression factor of the central fit, obtained at the starting scale $Q_0^2 = 1.9\,\text{GeV}^2$ are compared to variations of the model input parameters. The corresponding PDF uncertainties, resulting from the experimental uncertainties of the input datasets, are evaluated using the Hessian method. To investigate the model uncertainties, alternative fits are performed, though only the central value of the alternative fit is presented.

The variations of the heavy quark masses relevant in this analysis are presented in Figs. D.1 and D.2. Different starting scales Q_0^2 have been tested and are shown in Fig. D.3. The impact of restricting the HERA data to values above a threshold Q_{min}^2 has been investigated and the results are presented in Fig. D.4. These variations do not alter the results for the strange quark distribution or the suppression factor significantly, when compared to the PDF fit uncertainty.

For the investigation of the parametrization uncertainties, also the PDF uncertainties, evaluated using the Hessian method, are presented. The parametrization of the different quark and gluon distributions, used for the central fit (see Sect. 6.1), are multiplied with terms of the form $(1 + D_j x)$ or $(1 + E_j x^2)$, with j representing the parton flavours. Figure D.5 presents the distribution for the case of D_{d_v} being a free parameter. Figures D.6 and D.7 present the results of $D_{\bar{d}}$ or $E_{\bar{d}}$ as free parameters of the fit. Figures D.8 and D.9 show the distributions for $D_{\bar{s}}$ and $D_{\bar{s}}$. None of the added parameters result in a significant difference of the strangeness distribution or strangeness suppression factor.

© The Editor(s) (if applicable) and The Author(s), under exclusive license
to Springer Nature Switzerland AG 2020
S. K. Pflitsch, *Associated Production of W + Charm in 13 TeV
Proton-Proton Collisions Measured with CMS and Determination of the Strange Quark
Content of the Proton*, Springer Theses,
https://doi.org/10.1007/978-3-030-52762-4

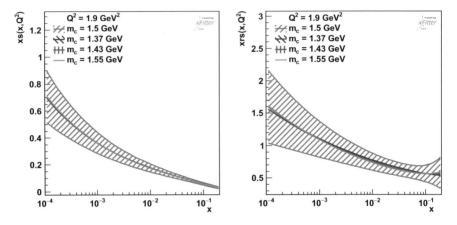

Fig. D.1 The strange quark distribution (left) and strangeness suppression factor (right) as a function of x, obtained at the starting scale $\mu_f = 1.9\,\text{GeV}^2$ for different values of the m_c parameter. The PDF uncertainties, obtained via the Hessian method, are presented for the central fit

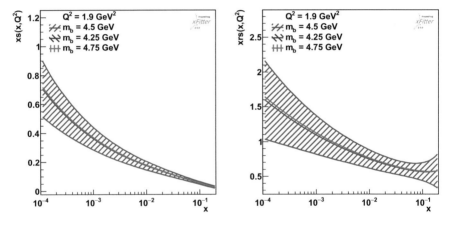

Fig. D.2 The strange quark distribution (left) and strangeness suppression factor (right) as a function of x, obtained at the starting scale $\mu_f = 1.9\,\text{GeV}^2$ for different values of the m_b parameter. The PDF uncertainties, obtained via the Hessian method, are presented for the central fit

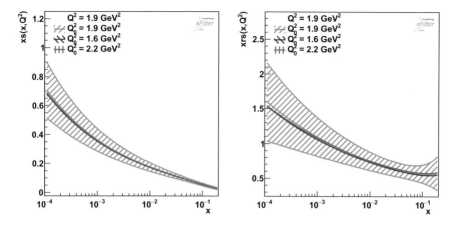

Fig. D.3 The strange quark distribution (left) and strangeness suppression factor (right) as a function of x, obtained at different starting scales. The PDF uncertainties, obtained via the Hessian method, are presented for the central fit

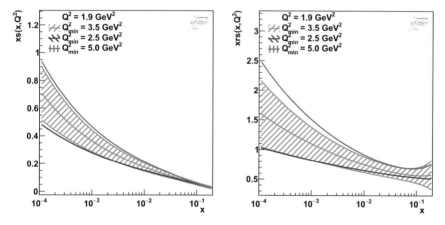

Fig. D.4 The strange quark distribution (left) and strangeness suppression factor (right) as a function of x, obtained at the starting scale $\mu_f = 1.9\,\mathrm{GeV}^2$ for different threshold values Q^2, applied to the HERA data. The PDF uncertainties, obtained via the Hessian method, are presented for the central fit

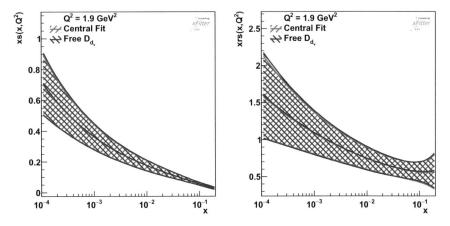

Fig. D.5 The strange quark distribution (left) and strangeness suppression factor (right) as a function of x, obtained at the starting scale $\mu_f = 1.9\,\text{GeV}^2$. The parametrization of the central fit is compared to a parametrization in which the D_{d_v} parameter is free. The PDF uncertainties are obtained via the Hessian method

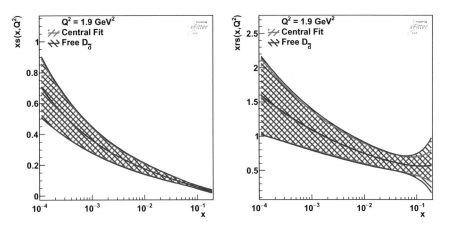

Fig. D.6 The strange quark distribution (left) and strangeness suppression factor (right) as a function of x, obtained at the starting scale $\mu_f = 1.9\,\text{GeV}^2$. The parametrization of the central fit is compared to a parametrization in which the $D_{\bar{d}}$ parameter is free. The PDF uncertainties are obtained via the Hessian method

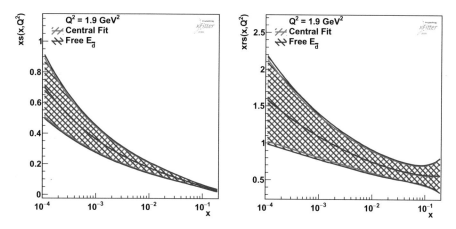

Fig. D.7 The strange quark distribution (left) and strangeness suppression factor (right) as a function of x, obtained at the starting scale $\mu_f = 1.9\,\mathrm{GeV}^2$. The parametrization of the central fit is compared to a parametrization in which the $E_{\bar{d}}$ parameter is free. The PDF uncertainties are obtained via the Hessian method

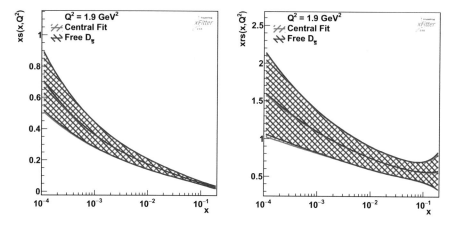

Fig. D.8 The strange quark distribution (left) and strangeness suppression factor (right) as a function of x, obtained at the starting scale $\mu_f = 1.9\,\mathrm{GeV}^2$. The parametrization of the central fit is compared to a parametrization in which the $D_{\bar{s}}$ parameter is free. The PDF uncertainties are obtained via the Hessian method

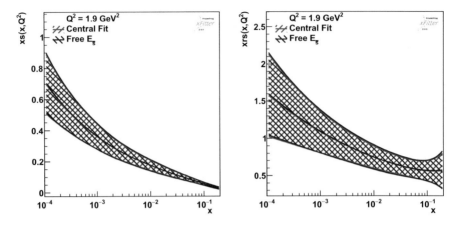

Fig. D.9 The strange quark distribution (left) and strangeness suppression factor (right) as a function of x, obtained at the starting scale $\mu_f = 1.9\,\text{GeV}^2$. The parametrization of the central fit is compared to a parametrization in which the $E_{\bar{s}}$ parameter is free. The PDF uncertainties are obtained via the Hessian method

Reference

1. The CMS Collaboration. CMS Luminosity Measurements for the 2016 Data Taking Period. CMS Physics Analysis Summary CMS-PAS-LUM-17-001 (2017) http://cds.cern.ch/record/2257069

Printed in the United States
by Baker & Taylor Publisher Services